PREHISTORIC Journey

A History of Life on Earth

Kirk R. Johnson and Richard K. Stucky

Denver Museum of Natural History
Roberts Rinehart Publishers

International Standard Book Number 1-57098-056-X, 1-57098-145-4 (PB)
Library of Congress Catalog Card Number 95-69271

Published by ROBERTS RINEHART PUBLISHERS
5455 Spine Road, Boulder, Colorado 80301

Published in the UK and in Ireland ROBERTS RINEHART PUBLISHERS
by Trinity House, Charleston Road, Dublin 6, Ireland

Distributed in the U.S. and Canada by Publishers Group West
Printed in Hong Kong

First Edition

THIS PUBLICATION WAS SUPPORTED IN PART
BY FUNDING FROM THE AMOCO CORPORATION

Project Manager Betsy R. Armstrong, assisted by Ann W. Douden
Production Ana Hill
Cover Painting John Gurche
Watercolor Paintings Greg Michaels
Photography Rick Wicker
Editor Alice Levine
Editorial Assistant James Alton
Proofreader Lori Kranz
Reviewers Donald Prothero, Occidental College;
Hermann Pfefferkorn, University of Pennsylvania; and
Kenneth Carpenter, Denver Museum of Natural History

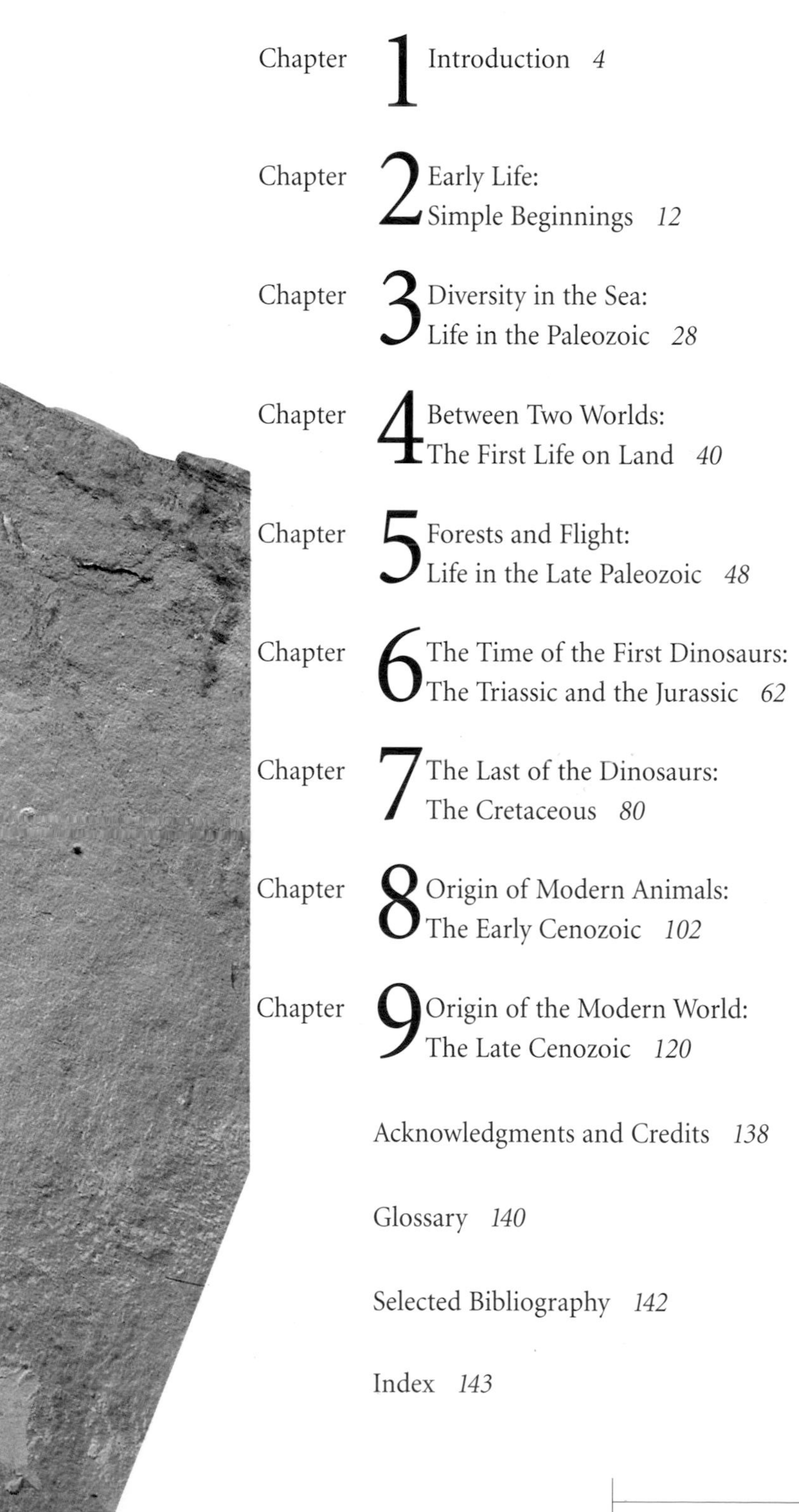

contents

CONTINENTS *on the move*

[600 million years ago to the present]

Forces within the earth cause the continents to move at speeds of one to a few inches a year. Over the last 600 million years, continents have split, collided, and finally arrived at their present configuration. Plants and animals are passengers on these drifting continents, and the history and evolution of species have been influenced strongly by continental position.

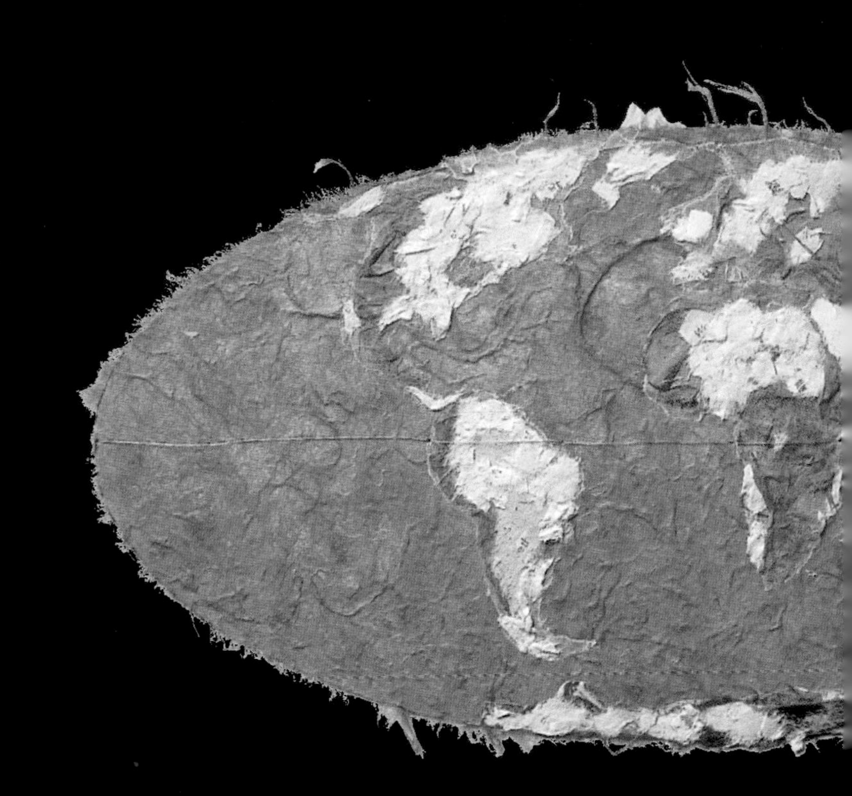

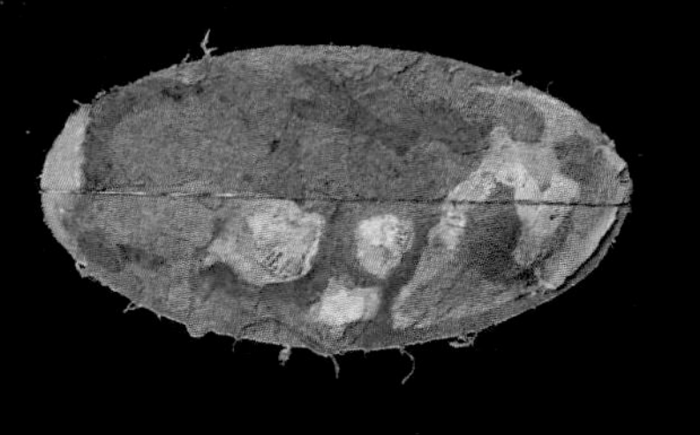

The Precambrian World
[600 MILLION YEARS AGO]

The continents are grouped in the Southern Hemisphere. Plankton and the earliest forms of multicellular animals inhabit the oceans but no plants or animals live on land.

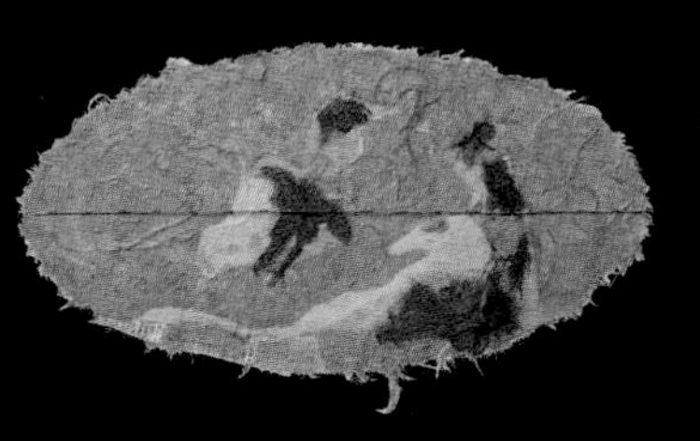

The Silurian World
[425 MILLION YEARS AGO]

The continent that will become North America has drifted across the equator and is largely covered by shallow seas.

The Devonian World
[395 MILLION YEARS AGO]

Continents begin to drift together and shallow seas cover portions of the landmasses. Plants, insects, and other arthropods colonize the land.

The Carboniferous World
[295 MILLION YEARS AGO]

Ice on the South Pole creates glaciers and affects global climate as the continents drift together to form the supercontinent known as Pangaea.

Today's World
[THE 20TH CENTURY]

The present configuration of continents and oceans emerges from millions of years of continental drift. Ice covers both poles, temperate forests and grasslands blanket the mid-latitudes, a belt of deserts rings the low latitudes, and rainforests and savannas occupy the tropics.

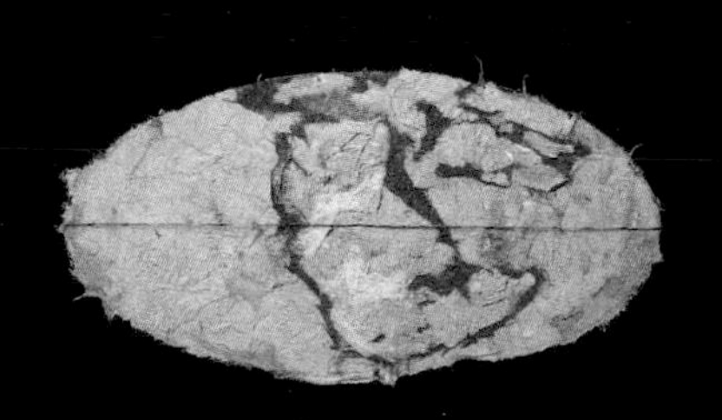

The Triassic World
[225 MILLION YEARS AGO]

The interior of the supercontinent Pangaea is alternately subjected to giant monsoonal rains and to dry periods.

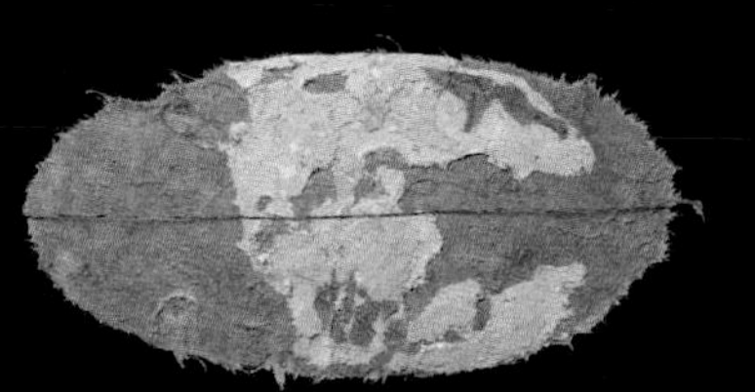

The Cretaceous World
[120 MILLION YEARS AGO]

The supercontinent Pangaea splits into two parts: Laurasia and Gondwana. Global climate is warm and many continents are flooded by shallow seas.

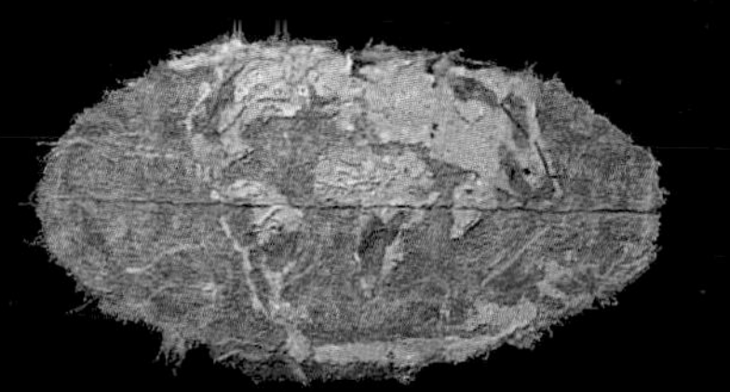

The Eocene World
[50 MILLION YEARS AGO]

A greenhouse atmosphere creates a warm earth with no polar ice caps. The modern continents are taking shape.

The Miocene World
[20 MILLION YEARS AGO]

An ice cap forms at the South Pole, the climate cools, and the continents are drifting toward their present locations.

1 chapter ONE

introduction

Today's World

[THE 20TH CENTURY]

The present configuration of continents and oceans emerges from millions of years of continental drift. The splitting and collision of the continents are responsible for the modern distribution of plant and animal species.

A fossil is evidence of ancient life. It can be as complete and telling as a whole frozen mammoth from 30,000-year-old Siberian permafrost or as incomplete and enigmatic as a burrow trace of an unknown 500-million-year-old marine worm. Fossils are windows into past worlds. They tell us stories that would otherwise go untold. Without fossils, we would know nothing of the dinosaurs, the first plants on land, or the vast diversity of life forms that once lived in the sea but are now extinct.

Through the study of fossils, paleontology, humans are able to reconstruct the origins and evolution of life and to reveal the history of our own species. *Prehistoric Journey* is our attempt to illustrate this incredible and fascinating saga. Our perspective is an ecological one. We view the history of life as a series of interlinked and ever-changing ecosystems filled with an incredible panoply of diverse organisms. Woven into the tapestry of life's past is the process of evolution, the mechanism that explains how organisms change over time into novel and different forms.

The methods of paleontology are as varied as the fossils themselves. Paleontology draws heavily on two main disciplines, geology and biology. Geology examines earth's history through analysis of the sediments and rocks that have been laid down through time. Biology explores the diversity of life and its special attributes such as anatomy, ecology, and behavior. Paleontology also relies on dozens of other scientific and technological disciplines. For example, engineering provides background information on the structure and form of skeletons, and physics provides the radioactive age constants of isotopic elements that allow us to date the age of rocks. The kit of the paleontologist contains tools as crude as ice picks, sledge hammers, and dynamite, and as sensitive and sophisticated as scanning electron microscopes and remote sensing satellites.

The search for the origins of life is an important and complicated one. It is made difficult by the immense spans of time involved. The earth formed 4.6 billion years ago, and the first traces of life appear in rocks that are about 3.5 billion years old. The process that produced single-celled life, then cells with a nucleus, and finally multicellular life took nearly 3 billion years. At that point life had evolved to forms that were just big enough to be seen with the naked eye. By 600 million years ago, the diversity of life had begun to leave its mark in the form of abundant fossils in the rocky pages of earth's history.

In order to read these pages, we need to understand the major forces that act upon the earth and the specific processes by which fossils are created. Planet earth is an active place. The heat within the planet causes continents to move across the surface at rates that are as rapid as inches a year. The fault zones within these continents and the friction at their margins cause volcanic eruptions, earthquakes, lava flows, and even the construction of whole mountain ranges. Energy from the sun affects the earth's air and water in a variety of interesting and active ways. At the equator, where the sun's rays are strongest, air is heated and rises to great elevations where it cools and releases torrential rainstorms. At the poles, where the sun's energy is weakest, water freezes into huge ice sheets and surging glaciers. The air currents in the atmosphere and the circulation of water in the oceans are driven by the rotation of the earth and the topography of its surfaces. The result is what we call weather. The earth's orbit around the sun produces the seasons. Changes in these patterns of air flow, ice formation, and water circulation as well as in the gaseous composition of the atmosphere have made an ever-changing climate one of earth's most characteristic features.

Wind, water, and weather take their toll on the land in the form of erosion. Exposed surfaces are subject to erosion, and in time rocks and mountains are reduced to gravel, sand, and mud. These sediments are carried down by rivers

Trilobite

Trilobites, among the most distinctive fossils, were common early in earth's history but became extinct more than 250 million years ago.

Trilobite

[Olenellus gilberti]

Early Cambrian, 530 mya
Pioche Formation, Nevada
Trilobite length: 5.3 inches, DMNH 6961

Pages of time
These thin layers of shale were deposited on the bottom of an ancient lake. The layers of shale are like pages of time. Paleontologists search these rocks for fossils of plants and animals that lived in and near the ancient lake system.

to lakes, seas, and oceans where they are deposited in layers. Other sediments are formed partially or entirely from plants or animals. Forested swamps accumulate peat composed of partially decayed plant debris while the bottoms of shallow warm seas become layered with the carbonate bodies of marine animals. In time, the layers are buried by more layers, and a tremendous thickness, miles of mud, sand, peat, and lime, accumulates. The immense weight of the overlying layers compresses the lower layers, squeezing out the water and air and transforming the sediment into sedimentary rock such as conglomerate, sandstone, shale, coal, and limestone. Many of these layers lie buried deep beneath the surface of the earth or below the murky seafloor but many are lifted back to the surface by the motions of the continents and the forces that drive mountain ranges up into the air. These uplifted layers, sometimes lying as flat as when they were deposited and sometimes folded and distorted by the process of being thrust back to the surface, are visible in the sides of hills and mountains, along sea coasts and river valleys, and in quarries, roadcuts, and construction sites.

Sedimentary rocks of various ages bear the fossils of those ages. Plants and animals that lived and died where sediments are accumulating often are buried by the sediments. If burial is rapid and the chemistry is right, the organism will be preserved as a fossil when the sediment becomes a rock. Usually it is only part of the organism that survives as a fossil. Clam fossils preserve the shells but rarely the soft body parts. Dinosaurs leave their bones, sometimes their skin impressions, but never their eyes or entrails. Plants fall apart as they live, leaves falling by the millions each fall, flowers and pollen in the spring, seeds and fruit when they ripen. So fossil plants are usually just parts of ancient plants, a leaf imprint here, a petrified log there.

Lotus leaf

[Nelumbo sp.]

Fossils of aquatic plants, like this lotus, are often found in the deposits of ancient lakes.

Middle Eocene, 48 mya
Green River Formation, Colorado
Slab length: 17.5 inches, DMNH 6946

Insects

[Diptera]

Usually rare as fossils, insects are commonly preserved in the fine-grained oil shales of the Green River Formation. These flies and the crane fly settled to the bottom of the ancient lake and were covered by fine mud.

Middle Eocene, 48 mya
Green River Formation, Utah
Slab length: 5.2 inches, DMNH 7840

Rhinoceros bone bed

Extensive bone beds are found at several sites in the western United States. They contain skeletons of both young and old animals, which suggests that the animals, like these rhinoceroses, lived in herds and met a catastrophic death.

Rhinoceros bone bed

[*Trigonias osborni*]

Late Eocene, 35 mya
White River Formation, Colorado
Slab length: 5.3 feet, DMNH 425

Understanding how ancient plants and animals looked and how they lived is a job for a detective. Paleontologists are the detectives of time, searching out the whos, hows, wheres, and whys of ecosystems long dead and buried. Since fossils are so often fragmentary, other clues become very important. What is the nature of the sedimentary rock that houses the fossil? What can be said about the environment in which the organism was buried? Does the fossil show any clues as to why the animal died? Was it chewed by scavengers before it was buried? Was the burial fast or slow? The answers to these types of questions can often be had by careful observation.

Certain fossil localities contain an unusual wealth of information. For a variety of reasons, ranging from the nature of the original ecosystem to the speed of its burial to the type of fossilization, these sites preserve more or better fossils than the average site. These sites have come to be known as *lagerstätten*, a German word that means "mother lode," because the sites are gold mines of information. Examples of well-known lagerstätten include the Burgess Shale of British Columbia and the Solnhofen Limestone of Germany. The Burgess preserves the bodies and soft body parts of over a hundred creatures that flourished very early in the evolution of life. The fine-grained limestone at Solnhofen has produced nearly perfect skeletons complete with wings and feathers of *Archaeopteryx*, the first known bird. In many lagerstätten, the variety of fossils and their high quality of preservation allow the fossil detective to gather far more clues and to make better and more accurate reconstructions of the ancient ecosystems.

Well-preserved fossils display the shape and anatomy of long-extinct animals. Careful analysis of the morphology of a fossil and comparisons with its nearest relatives, living and dead, is the way that we begin to understand how an animal lived. The type of teeth suggests diet; the size of muscle scars speaks to the mode of locomotion and the movement of the limbs. In plants, the size of the fossil leaves reflects the annual rainfall while the spacing of fossil tree trunks stands as a mute reminder of the ecology of the extinct forest. The behavior of whole groups of animals or evidence that they lived in herds may be preserved in bone beds or fossil trackways. Even features that no longer occur, such as the spiked tail of the stegosaur or the 9-inch-long canine teeth of the saber-toothed cats, can be understood when interpreted in an evolutionary and adaptive context. And slowly, the ecology of ancient worlds begins to emerge. Evolutionary theory predicts that both extrinsic factors, such as climate change and extraterrestrial impact, and species interactions will have dramatic effects on the adaptations of life forms.

In this book, we look at the history of life by describing and resurrecting eight exquisite fossil sites. Each chapter of this book begins with a description of a fossil locality as it appeared when it was alive and functioning. These reconstructions represent our best attempt at a series of snapshots through time—progress reports on the evolution of the planet and the life forms on it. We have visited these sites, collected the fossils, and studied the geologic context. The reconstructions represent the collaboration of paleontologists, biologists, geologists, and artists. We are intrigued by fossils and the stories they tell and have attempted to illustrate this Prehistoric Journey with images of fossils that capture the beauty and mystery of prehistory.

Ediacara Hills, Australia

[620 Million Years Ago]

Blue light illuminates a rippled sandy seafloor above which a light surf breaks. Several bizarre organisms litter the sandy seafloor. Translucent Ediacaria, Mawsonites, *and* Kimberella *lie on the seafloor. They are shaped like disks and range from 3 to 8 inches in diameter. Green and blue blade-shaped* Charniodiscus *rise 2 to 4 feet above their rooted holdfasts in the sand and bend gently in the moving water. Shaped like placemats, oval* Dickinsonia *and* Spriggina, *some up to 25 inches long, rest on the seafloor. All of these are animals with no mouths or stomachs. They obtain critical nutrients by absorbing them either directly from seawater or with the aid of photosynthetic algae that live within their body tissues. There is little interaction between them, and no predators threaten prey in this sea. It's a peaceful kingdom but not a particularly active one.*

chapter TWO 2

Early Life:

Simple Beginnings

Ediacara Hills, Australia

[Today]

The Ediacara Hills are located in the dry desert scrub between the dune fields of Lake Torrens and the Flinders Ranges in South Australia. Today, kangaroos and emus wander across the ridges of red sandstone that mark the site where the Rawnsley Quartzite is exposed. This formation became famous in 1947 when Reg Sprigg, an Australian geologist, discovered what were then the oldest known fossils in the world. The Rawnsley Quartzite originally was deposited as sediment in a shallow sea that covered Australia 600 million years ago. The slabs of sandstone in the Ediacara Hills bear ripple marks that were made by storm currents moving the sand on the seafloor. Imprinted on the undersides of these slabs are the varied patterns of the animals that have come to be known as the Ediacaran Fauna. Variously shaped like coins, coasters, paddles, and placemats, these enigmatic early animals remain the center of a heated debate on the nature of the first multicellular animals on earth. One group of scientists maintains that these fossils represent the first occurrences of such well-known groups as the jellyfish, sea pens, arthropods, and flatworms, while other scientists consider them a completely extinct group with no known descendants and have named them the Vendozoans.

Vendozoan

[Dickinsonia costata]

Faint imprints on red sandstone are all that remain of the enigmatic animals of Ediacara. Blade-like and leaf-shaped, this animal probably absorbed nutrients through its surface or incorporated photosynthetic algae into its tissue.

Late Precambrian, 600 mya
Pound Quartzite, Australia
Animal length: 1.8 inches, DMNH 6015

2 chapter TWO

Early Life:

Simple Beginnings

The Precambrian World

[600 MILLION YEARS AGO]

The continents are grouped in the Southern Hemisphere. Plankton and the earliest forms of multicellular animals inhabit the oceans but no plants or animals live on land.

The planet becomes livable

Our exploration of early life must begin with the earth before life appeared. All known living organisms are on earth, but there was a time when nothing lived here. Our solar system and its planets formed from a huge swarm of interstellar debris as thousands of asteroids were pulled together by gravity. The newborn earth was a galactic punching bag, bombarded by a multitude of meteorites. Radioactivity heated the earth from within and allowed heavier elements to concentrate at its core. Thus, the earth has a core of iron and nickel, and a mantle and crust of lighter elements. The intensely cratered surface of the moon, unaltered by erosion because the moon has no atmosphere, records this early period of intense bombardment. Life could not exist on the young earth.

As the bombardment decreased about 4 billion years ago, the temperature of the earth cooled and a primitive atmosphere formed around it. If the earth had simply cooled from a mass that was molten nearly 4 billion years ago, it would be a solid cold ball. The earth today has a hot and partially molten core because its rocks include radioactive elements that decay to stable elements and give off heat in the process. The same type of energy that powers nuclear reactors also keeps the earth from cooling down, and the process of radioactive decay allows us to date important events in earth's history.

In addition to having a comfortable temperature, earth provided several other essential components for the first life forms. Oxygen is critical for many forms of life today, but the earth's early atmosphere contained very little free oxygen. Instead, the atmosphere was probably composed of nitrogen, hydrogen, carbon dioxide, ammonia, and methane. Today the earth's atmosphere is composed primarily of nitrogen (79 percent) and oxygen (21 percent), with trace amounts of other gases such as carbon dioxide. Animals require oxygen to breathe, so early life must have been of a type that did not need oxygen to survive. Such organisms live today and are called anaerobic bacteria. They can be found either deep beneath the surface of the earth, in hot springs like those at Yellowstone National Park, or in the mud at the bottom of the ocean. These anaerobic organisms are single-celled and have no nucleus. They provide us with clues to understanding the first organisms. But where did they come from?

Although we are not quite sure of the answer to that question, we do know from water-lain sedimentary rocks in Greenland that 3.8 billion years ago earth had free-standing water. In addition, chemicals essential to life existed on the surface of the earth. Thus, the early world became the laboratory in which the elements of life combined in water to form the first living organism. In order to understand how this happened, scientists are trying to re-create the conditions of the early earth and to synthesize life in the laboratory. They have succeeded in making amino acids, the basic components of proteins. DNA (deoxyribonucleic acid) and RNA (ribonucleic acid) molecules, which are composed of complex arrays of amino acids and are the templates for all living organisms, have yet to be artificially created.

All life forms that we know of are composed of the same organic compounds: mixtures of carbon, hydrogen, oxygen, nitrogen, sulfur, phosphorus, and other chemical elements. The first life forms were microscopic single-celled organisms that shared several characteristics with all organisms that have ever existed: They consumed energy, they grew, and they reproduced. As oxygen increased in the atmosphere, more complicated single-celled organisms, each with their own nucleus, appeared. It was not until 600 million years ago that multicellular animals began to leave traces as fossils in the pages of earth's history.

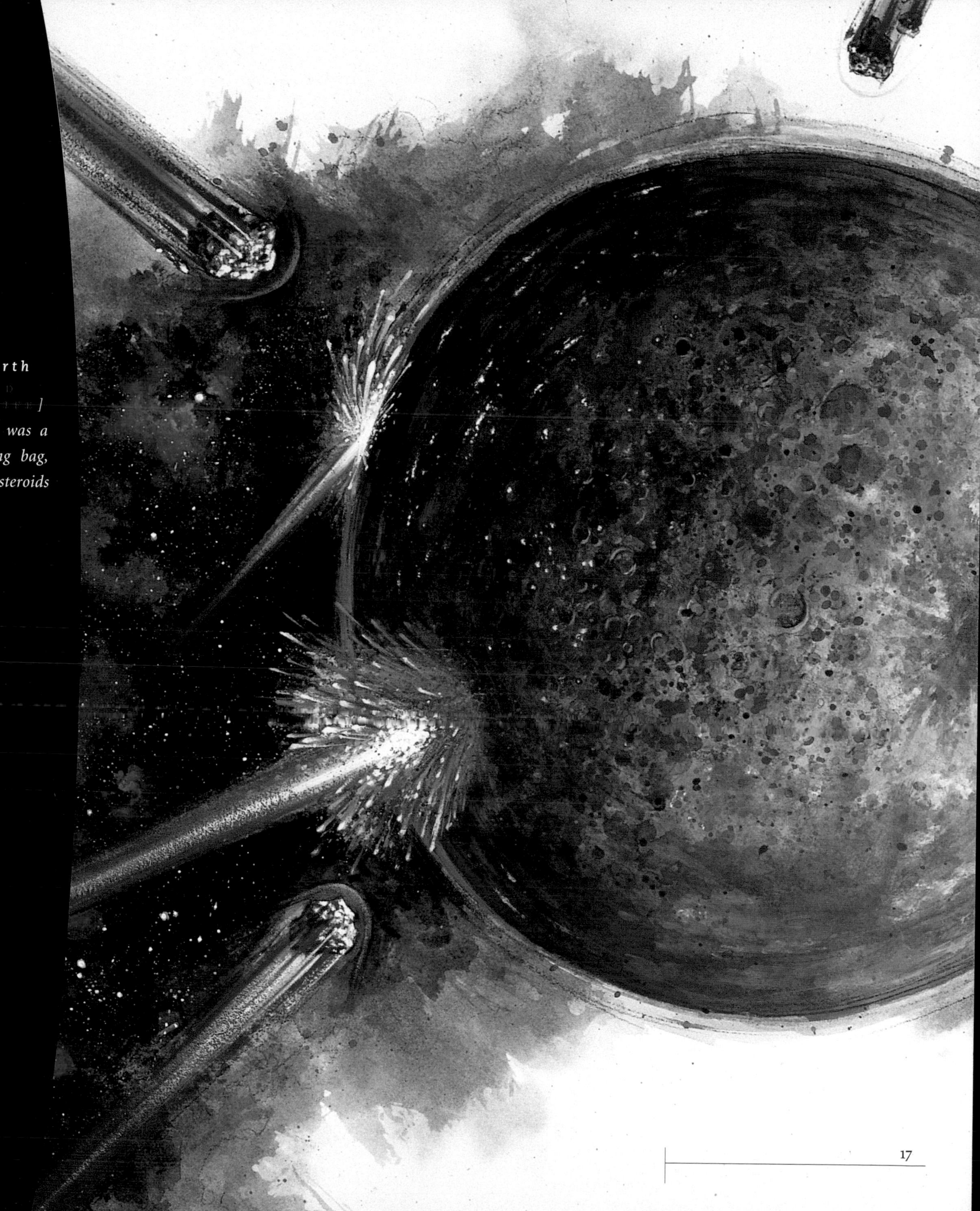

Hadean earth

[A WORLD WITHOUT LIFE]

The early earth was a galactic punching bag, pummeled by asteroids and meteorites.

Chemical soup

The early oceans contained the water and chemical compounds that were the components of the first forms of life.

Earliest rocks on earth

Our earth and its moon were both formed about 4.6 billion years ago. On earth, erosion by wind and water wears rocks away to sediment, and the process of plate tectonics causes the edges of the oceanic plates to be shoved under the continental plates and destroyed by melting. Thus, most rocks that formed early in the earth's history do not exist today. They have long ago been eroded or melted and reconstituted as younger rocks. However, on the moon, which lacks plate-tectonic energy and the destructive atmosphere and water of the earth, the rocks record their antiquity. Thus, we can indirectly determine the age of the earth by dating moon rocks and meteorites.

The oldest known rocks on earth, which come from the Slave Province of Canada's Northwest Territories, are about 4 billion years old. But because these rocks are metamorphic, sedimentary rocks that have been cooked to the point of recrystallization, their fossils have been destroyed. The oldest rocks that preserve sedimentary structures are around 3.8 billion years old and are located at Isua on the north coast of Greenland. They have not been treated well by time and have been heated past the point where fossils could have been preserved. The oldest sedimentary rocks that have survived to the present in a relatively intact form are 3.5 billion years old and are found in northwestern Australia and in southwestern Africa.

World's oldest rock

[Acasta Gneiss]

The oldest rocks in the world occur in Canada's Northwest Territories. This piece of Acasta Gneiss comes from an outcrop that has been radiometrically dated to 3.92 billion years.

Earliest life

[MOLECULES TO CELLS]

Chemical compounds combined in increasingly complicated forms. Life as we know it began when the first living single-celled microorganisms started to consume energy, grow, and reproduce. Strands of photosynthetic cyanobacteria grouped together to form the first ecosystems.

The fossil record of early life

Three independent lines of evidence support the existence of life on earth 3.5 billion years ago, and there is a strong possibility that life existed even earlier. The lines of evidence are the microfossils themselves, the carbon-isotope record, and the rock record of oxygen.

The fossil record of earliest life is difficult to read because early life forms were very small and nondistinct. Also, there are few rocks of great antiquity left on the planet. Until the 1950s, paleontologists were unsuccessful in locating fossils of very early life forms from the time known as the Precambrian. Though representing 90 percent of the earth's history, Precambrian rocks appeared to be barren of fossils.

One of the reasons that Precambrian rocks seemed to contain no fossils was that paleontologists looked only with their eyes and not with microscopes. This situation changed in 1953 when Elso Barghoorn, a Harvard paleobotanist, used a technique called thin sectioning to grind pieces of 2-billion-year-old chert from Ontario into translucent slices. Under the microscope, these thin sections showed cross sections of microscopic cells that had been replaced by silica. This technique is a very valuable tool in the search for the world's oldest fossils.

The search for earliest life has now extended to the oldest sedimentary rocks in the world, the 3.5-billion-year-old Warrawoona Group of geologic formations in northwestern Australia and the similarly ancient Swaziland Group in southwestern Africa. Amazingly, these ancient rocks preserve fossils of once-living organisms. They contain evidence of life in three forms: fossils of microscopic bacteria, sedimentary

A slice of stromatolites

[Collenia columnaris]

The red concentric bands in this stromatolite were formed as photosynthetic bacteria and algae trapped particles of mud at the edge of a Precambrian sea.

Precambrian, 2,000 mya
Biwibak Formation, Minnesota
Slab length: 8.3 inches, DMNH 7108

structures known as stromatolites, and a peculiar organic chemistry. Thin sections of the Warrawoona rocks reveal a diversity of microbial fossils, including elongate strand-like cells that are extremely similar to living cyanobacteria. One site in northwestern Australia has produced eleven different types of microbes. This high diversity strongly suggests that life existed on the planet well before even these ancient rocks were deposited.

Today, stromatolites, mounds the size and shape of upside-down bushel baskets, are formed as a slimy layer cake of blue-green bacterial scum and muddy sediment at the edge of the sea. The blue-green bacteria and other microbes that form modern stromatolites gain their energy from the sun through a process known as photosynthesis. They produce carbohydrates and oxygen by combining the energy of the sun with carbon dioxide and water. The organisms use the carbohydrates to grow, and the oxygen is released as a waste product of the reaction. The stromatolites get larger as mud particles in the seawater are trapped in the sticky bacterial filaments. The bacteria grow around and through the mud, and a new layer forms. Stromatolites exist today in a number of shallow marine areas, the most famous being Shark Bay in West Australia. Rocky outcrops in Australia and Africa have preserved stromatolites that are 3.5 billion years old, which suggests that even the earliest known life forms may have been photosynthetic.

Geochemistry, the study of chemical compounds in rock, provides an independent check for the presence of life. There are two stable carbon isotopes: Carbon-12 and Carbon-13. Living organisms more often utilize Carbon-12, the lighter isotope, whereas deposits of nonorganic carbon are enriched in Carbon-13, the heavier isotope. An enrichment of the lighter isotope in ancient rocks is therefore a test for the presence of life. Indeed, lighter isotopes predominate in rocks that are 3.5 billion years old and have even been measured in the rocks from Isua, Greenland, that are 3.8 billion years old.

The history of the waxing and waning of atmospheric oxygen is also recorded in the earth's rock record. Remember that the early atmosphere had very little oxygen and that photosynthesis produces oxygen as a waste product. So even at the beginning, life began to alter the nature of its environment. Oxygen in the early atmosphere had a number of potential fates: It could react with volcanic gases; it could be used by oxygen-loving microbes; and it could combine with iron to form rust. These three factors combined to keep oxygen from accumulating in the atmosphere. Some of the most visible and convincing evidence for the low levels of oxygen in the early atmosphere can be seen in the Mesabi Range of northern Minnesota where a thick sequence of early Precambrian rocks known as banded iron formations is exposed. Composed of banded layers of iron oxides, these rocks formed when oxygen combined with iron and settled to the bottom of ancient seas as a rain of tiny rust particles. Banded iron formations occur around the world in formations that were deposited between 3.5 and 2 billion years ago. Early life was rusting the planet.

About 2 billion years ago, most of the available iron had reacted with oxygen, and the creation of banded iron formations ceased. Because there was no longer any iron to react with, oxygen began to build up in the atmosphere. The stage was set for animals that breathed oxygen. Ironically, the increase in oxygen probably caused the extinction of some microbes that had evolved in a world without oxygen.

The first evidence of eukaryotes, organisms with cells that have a nucleus, is found in early Proterozoic rocks (rocks deposited between 2.5 billion and 545 million years ago). Today eukaryotes include all plants, animals, fungi, and a variety of algae. Prior to 2.5 billion years ago, all organisms were prokaryotes, single-celled microbes whose DNA molecules were loosely contained inside the cell membrane. Eukaryotes contain their DNA in a membrane-bounded nucleus within the cell. Eukaryotes also contain a number of other specialized units within their cells. Plant cells, for example, contain discrete chloroplast organelles that perform the function of photosynthesis for the cell. Lynn Margulis, a University of Massachusetts biologist, suggested that these complex cells first evolved when a prokaryote cell engulfed another similar cell and they began to function together, becoming a more complicated organism. This theory is now widely accepted.

The early eukaryotes, like the prokaryotes, reproduced by asexual cell division. About 1.2 billion years ago, the eukaryotes began to reproduce sexually and the world has never been the same since. In sexual reproduction, an individual combines one-half of its genetic blueprint, the DNA, with that of another individual, creating an offspring that is a composite of the two. Sexual reproduction allows a rapid mixing of genetic material. A burst of evolution followed the appearance of sexual reproduction and the seas of the increasingly oxygen-rich world filled with a diversity of marine plankton.

The bloom of photosynthetic organisms that appeared around 1 billion years ago was a juggernaut that generated more oxygen from atmospheric carbon dioxide. Carbon dioxide is known as a greenhouse gas because even in small atmospheric concentrations, it serves to insulate the world and to raise its temperature. About 600 million years ago, a severe ice age occurred. It is possible that this climatic cooling was, in part, caused by the biological greed for atmospheric carbon dioxide.

Ediacara and the vendozoans

About 600 million years ago, the first animals large enough to be seen without magnification appeared on earth. These organisms, originally found in the Ediacara Hills in South Australia, are preserved as impressions on the undersides of layers of sandstone that were deposited by storms in a shallow sea. The impressions are round, oblong, and spindle-shaped and are marked with numerous radiating striations. Some were huge by Precambrian standards, as large as a doormat. The scientists who first saw these fossils interpreted them to be the ancestors of modern marine groups. They classified the fossils as jellyfish, flatworms, sea pens, and arthropods.

In the fifty years since the discovery at Ediacara, similar faunas have been found in Europe, Asia, Africa, and North America. In 1983, Adolf Seilacher, a brilliant German paleontologist, suggested that the Ediacaran organisms were an entirely novel type of organism, not the ancestors of the living groups and not animals as we know them. He pointed out that modern jellyfish and sea pens leave very indistinct impressions on sand and that those impressions are rare in rocks younger than the Precambrian, whereas the Ediacara traces are very distinct and common. He named these Ediacaran animals vendozoans and argued that they became extinct at the beginning of the Cambrian when the earth's first predators evolved. Seilacher described these animals as having a quilted construction like an air mattress, with fluid-filled compartments separated by rigid panels. This type of animal would absorb its food through its surface rather than consuming it through a mouth. An animal like this could be very successful in a world with no predators, where it would only have to be concerned with maximizing its surface area, not running or hiding from a threat. He suggested that these animals incorporated algae into their tissues and lived off the by-products of their photosynthesis like some modern

Vendozoan

[Dickinsonia costata]

This fossil is a trace of one of the first metazoans.

Late Precambrian, 600 mya
Pound Quartzite, Australia
Animal length: 1.8 inches, DMNH 6015

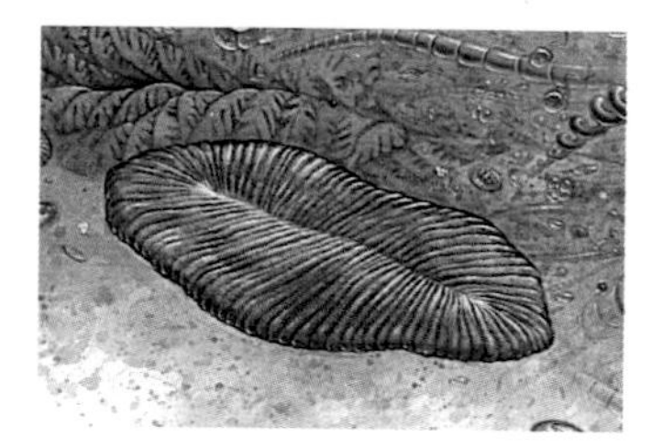

Photosymbionts?

[Dickinsonia costata]

The vendozoans may have incorporated photosynthetic algae in their bodies, giving them a green color.

reef animals, such as soft corals and giant clams, do today. This may explain the leaf-like and blade-like shapes of some of these oddballs. Modern leaves are flat and blade-like to present a maximum of photosynthetic tissue to the sun with a minimum of leaf volume. Perhaps these Ediacaran animals were like many green leaves growing on the seafloor. The fossils themselves preserve enough details of the animals to provoke the controversy about their biological relations but not to solve it, and the true nature of the world's oldest animals remains a mystery.

Cambrian explosion

[ORIGIN OF MARINE PHYLA]

Within a very short time about 545 million years ago, most existing major groups of marine animals evolved. The Cambrian explosion of life was the biological big bang.

The Cambrian explosion of life

The early Cambrian period, which began about 545 million years ago, has long been known for the earliest widespread occurrence of fossils and, therefore, by the name Cambrian Explosion. Animals with shells first appeared at that time. Cambrian rocks occur on all continents, and many sites preserve abundant shelly fossils. Because animals with shells are much more easily fossilized than animals with only soft parts, there has been a great deal of argument over whether the Cambrian represents the first big

Carpoid

[Castericystis vali]

This is one of the first echinoderms, relatives of modern sand dollars and sea urchins.

Middle Cambrian, 520 mya
Marjum Formation, Utah
Animal length: 2.2 inches, DMNH 6287

diversification of animals or merely the first big diversification of animals with shells. Cambrian rocks also contain many burrows and other traces, evidence that a variety of newly evolved animals were living in the seafloor sediment, mining it for nutrients, and perhaps hiding from newly evolved predators.

Stromatolites existed and even diversified in the Cambrian, but the bacterial scum that forms stromatolites is fast food for grazing marine herbivores such as snails. The appearance of a variety of herbivorous animals in the Cambrian ended the heyday of the stromatolites; they have been very rare ever since. The stromatolites that have survived to the present live in areas that are too salty or too hot for other marine animals.

The very first shelled animals in the Cambrian are represented by fossils of tiny enigmatic tube-like and disk-shaped shells. These were rapidly replaced by archaeocyathans, cone-shaped shells that formed some of the first reefs. But evolution in the Cambrian was rapid, and within a few million years, the world was populated with many of the groups that are still alive today, such as the mollusks, arthropods, sponges, corals, echinoderms, chordates, and worms. The most familiar fossils of the Cambrian are the bug-like trilobites, extinct relatives of crustaceans and insects. Most trilobites were smaller than 1 inch, but some individuals were much larger, such as the giant 18-inch *Acadoparadoxides*.

Paleontologists have discovered several Cambrian fossil sites that preserved not only fossil shells but also impressions of soft parts. The first discovered and most famous of these sites is the Burgess Shale in southern British Columbia. Sites have also been found in south China, northern Greenland, Utah, and other places.

The Burgess Shale preserves over 120 types of animals, representing many of the 33 phyla, or major groups, of animals that inhabit today's world. Interestingly, it also preserves a great variety of animals that do not appear to fit into any living phylum. Rather than starting small and expanding, it appears that the Cambrian Explosion gave life a big start. Stephen Jay Gould, the well-known Harvard paleontologist, has described this period as a time of great evolutionary experimentation. A great many body plans and organisms appeared, and only some of them survived to populate the modern oceans. The Burgess Shale and other similar Cambrian localities preserve the fossils of the first big predators. Perhaps the most intimidating animal was *Anomalocaris*, a creature up to 2 feet long that was armed with spiny feeding appendages and a formidable jaw composed of seven sharp blades. The arms race between predator and prey had begun.

Brachiopods

[Lingulella ampla]

Tiny oval shells of brachiopods are some of the most common Cambrian fossils.

Late Cambrian, 515 mya
Eau Claire Sandstone, Wisconsin
Slab length: 3.5 inches, DMNH 5075

Hyolithid

[Hyolithes cecrops]

Hyolithids were primitive mollusks, the group that includes clams and snails.

Middle Cambrian, 525 mya
Spence Shale, Utah
Shell length: 2.4 inches, DMNH 5369

Eocrinoid

[Gogia spiralis]

Eocrinoids were tiny echinoderms and relatives of the elaborate crinoids or sea lilies that populated shallow seas later in the Paleozoic Era.

Middle Cambrian, 520 mya
Wheeler Shale, Utah
Animal length: 2 inches, DMNH 5570

Trilobites

[COMMONERS OF THE CAMBRIAN SEAS]

Bug-like and beautiful, trilobites are treasured fossils that can be found in Cambrian rocks on every continent.

Trilobite

[*Olenellus gilberti*]

Early Cambrian, 530 mya
Pioche Formation, Nevada
Trilobite length: 2.2 inches, DMNH 8228

Trilobite

[*Alokistocare harrisi*]

Middle Cambrian, 520 mya
Wheeler Shale, Utah
Trilobite length: 1.9 inches, DMNH 5076

Trilobite

[*Acadoparadoxides briareus*]

Early Cambrian, 530 mya
Jbel Wawrmast Formation, Morocco
Trilobite length: 18 inches, DMNH 6171

Racine, Wisconsin

[425 Million Years Ago]

Sunlight filters through warm tropical seawater, enlivening a colorful reef. The crest of the reef is studded with purple, red, and blue coral heads. Calcium-secreting algae, mound-shaped stromatoporoids, and encrusting bryozoans are all part of the framework of this reef. Multitudes of clam-like brachiopods cluster in nooks and crannies. Like the corals, the brachiopods are varied shapes, sizes, and colors. Below the coral-studded reef crest is a crinoid meadow that looks like a submarine field of flowers. Crawling over the corals and brachiopods and between the crinoids are bug-like trilobites and coiled snails. A variety of carnivorous cephalopods (the group that includes squid and octopus) float menacingly above the reef as they stalk the trilobites. These cephalopods are distinguished by their shells, some of which are ribbed and the size and shape of baseball bats; others are partially coiled like cornucopias or fully coiled like nautiluses.

chapter THREE 3

Diversity in the sea:

Life in the Paleozoic

Today, the remains of an ancient reef are found in this flooded quarry in Racine County, just south of Milwaukee, Wisconsin. In fact, the whole southwestern shore of Lake Michigan is underlain by rocks composed of old Silurian reefs. The first outcrop ever recognized as a fossil reef in this area is now behind a grocery store in suburban Milwaukee; another makes a wooded hill near Milwaukee's venerable Country Stadium. As you drive west from Chicago on Interstate 80 through Cook County, the wooded countryside that flanks the highway drops away into a huge hole in the ground known as the Thornton Quarry. For over a hundred years, the city of Chicago has constructed highways, roads, and embankments from materials found in this quarry that were originally formed as tropical reefs.

How do we know these quarries are fossil reefs? If we look carefully at them, we see typical characteristics: fossils of marine animals, layering of rocks in a pattern similar to modern reefs, and the ubiquitous presence of carbonate rocks. Despite the vast exposures made by the quarries, it is difficult to study these reefs because the limestone has been altered to dolomite, a calcium magnesium mineral. The process of dolomitization occurs over time and wreaks havoc on the original limestone and its enclosed fossils. Many of the fossils are now ghostly outlines of their once prominent shapes. The core of the reef is made of solid blocky dolomite. Here, only the corals, stromatoporoids, and other "reef-builders" are preserved. The sloping flanks of the reefs are composed of flat dolomite layers, a few inches thick. The flank beds are remnants of the limy seafloor that surrounded the reef's core. They are composed entirely of crinoid debris, the broken stems and feeding arms of millions of the long dead crinoids. Although crinoids are commonly known as sea lilies, they are not plants. They are animals (echinoderms) whose cup-like bodies and lacy feeding arms are supported on long stem-like bases. They even have holdfasts that look like roots.

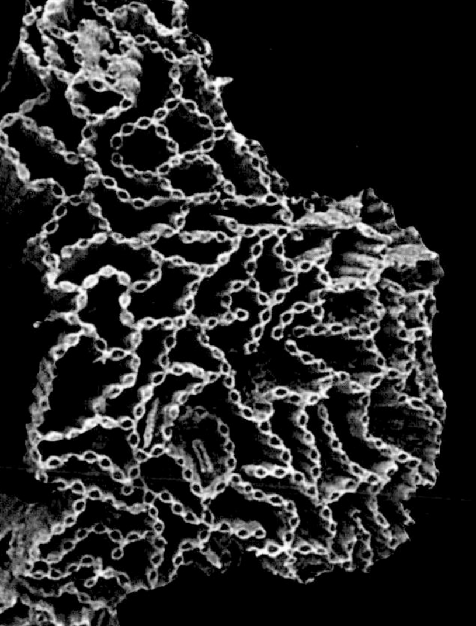

Chain coral

[*Halysites catenularia*]

The delicate filigree of this chain coral belies its age of 425 million years. Corals were common in the cores of the ancient reefs of the Silurian period.

Early Silurian, 425 mya
Racine Formation, Wisconsin
Coral length: 5.9 inches, DMNH 1831

Cephalopods

[*Dawsonoceras annulatum*]

Looking like ribbed pipes, these dolomite casts are the remains of the shells of cephalopods, predators of the Paleozoic seas.

Early Silurian, 425 mya
Racine Formation, Illinois
Slab length: 22 inches

3

chapter THREE

Diversity in the sea:

Life in the Paleozoic

The Silurian World

[425 MILLION YEARS AGO]

The continent that will become North America has drifted across the equator and is largely covered by shallow seas.

Lifestyles of Paleozoic marine animals

The animals and plants that evolved during the explosion of life in the Cambrian had only a few different lifestyles. We classify groups of animals that share similar lifestyles into "guilds" much like the tradesmen of medieval Europe were organized into guilds of carpenters, blacksmiths, and clockmakers. The most common guilds of Cambrian animals were grazers and sediment feeders. The grazers, such as trilobites, cruised the seafloor, consuming organic debris and algae. Sediment feeders such as marine worms would tunnel shallow burrows in the seafloor to hunt for bits of organic nutrient. Other guilds included some of the early small echinoderms that were able to filter plankton out of the water, a few inches above the seafloor. Monsters such as *Anomalocaris* notwithstanding, predators were relatively rare.

During the early Ordovician period, around 500 million years ago, animal life in the seas evolved into increasingly diverse communities. This burst of evolution caused a threefold increase in species over the late Cambrian. These new animals have been called the Paleozoic evolutionary fauna because they represent the lineages that were to dominate the seas for the next 250 million years until the Permian mass extinction decimated them and ended the Paleozoic era. Animals that appeared or greatly diversified during the Ordovician period include articulate brachiopods, crinoids and other echinoderms, cephalopods, bryozoans, graptolites, corals, and ostracods.

The new animals formed new guilds as they developed increasingly complex methods of exploiting food resources. The shelled cephalopods became dangerous swimming predators, able to cruise over the seafloor and snatch unwary prey. The graptolites, enigmatic beasts composed of radiating feathery arms, floated in the sea and filtered plankton. But the real diversification of lifestyles occurred on the seafloor. The Cambrian animals had confined their activities to within a few inches of the seafloor, but the creatures of the Ordovician began to exploit much more vertical space. The flower-like crinoids grew stems that extended their feeding arms as much as 4 feet above the bottom. Smaller echinoderms occupied the lower levels. Other animals made complex feeding burrows that extended deep into the seafloor sediments. This increasing complexity of feeding strategies allowed these animals to exploit a greater variety of food resources and created a much more complex series of ecosystems than had ever existed on earth.

Not all of the burrowing into the seafloor was for food. As predators, such as the swimming cephalopods, became larger and more numerous, animals began to seek refuge in burrows. Thus, we see how a trait that is useful for one purpose—feeding—sometimes becomes very useful for a completely unrelated purpose—hiding. Evolutionary theorists use the word *exaptation* to describe this process because the word *adaptation* does not adequately explain the serendipitous nature of the process. Exaptation, the exploitation of lucky traits, may be responsible for kick-starting some of the major diversifications in earth's history. The animals that found refuge by burrowing had stumbled upon a very successful technique. It took more than 100 million years before predators developed the ability to burrow beneath the seafloor.

The vertical partitioning of crinoids above the seafloor is analogous to the various vegetation levels in a modern forest. The organisms that live at certain levels do so because their particular traits allow them to be most efficient at those levels. Another method for exploiting vertical space in the water column is the formation of reefs. Reefs are organic structures that form on the floors of warm tropical seas.

Crinoids

This meadow of crinoids, or sea lilies, was buried in mud by a maritime storm. Many of the crinoids still preserve fossils of other animals, starfish and snails, that were feeding on the crinoid waste products when the storm hit.

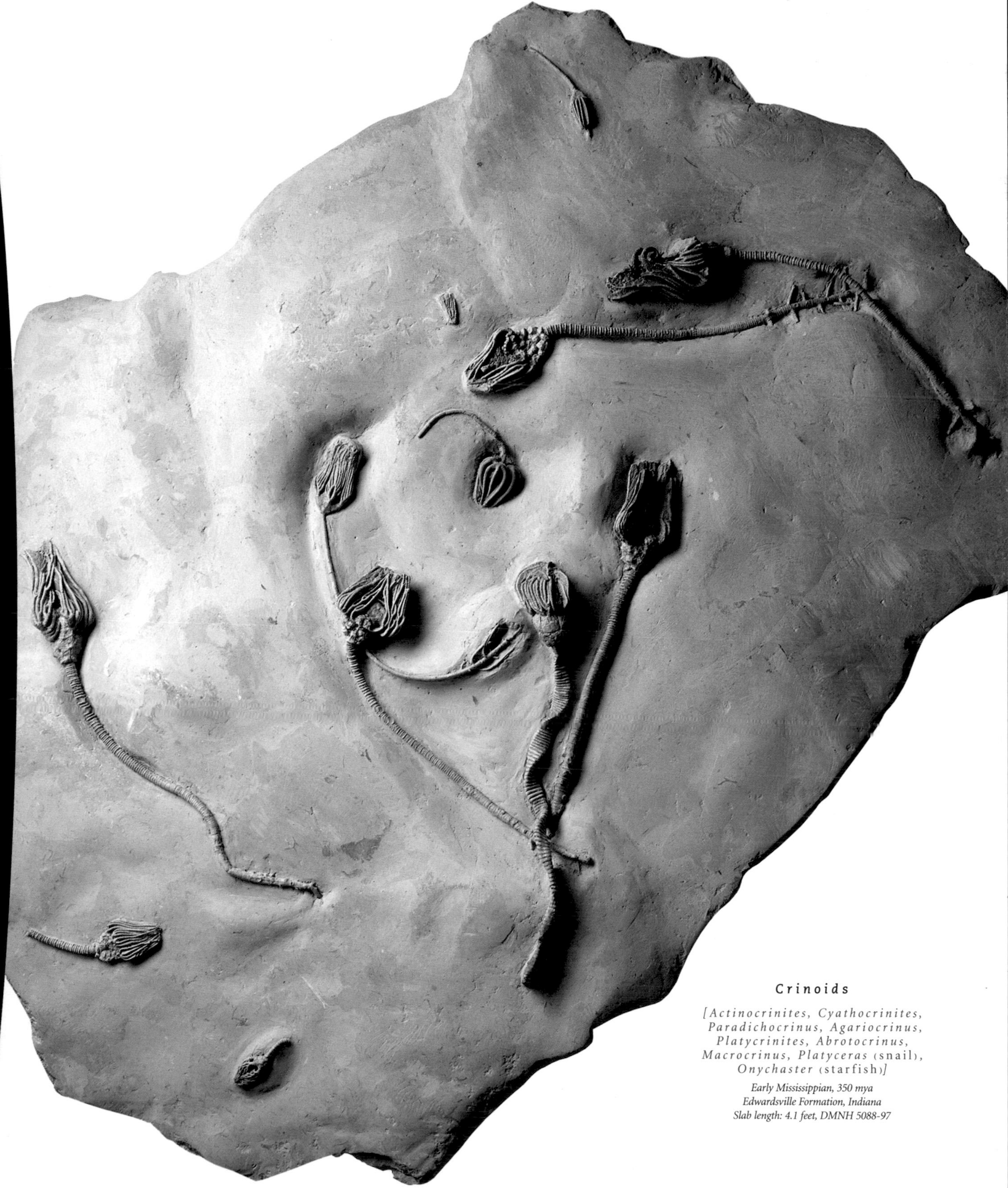

Crinoids

[Actinocrinites, Cyathocrinites, Paradichocrinus, Agariocrinus, Platycrinites, Abrotocrinus, Macrocrinus, Platyceras (snail), Onychaster (starfish)]

Early Mississippian, 350 mya
Edwardsville Formation, Indiana
Slab length: 4.1 feet, DMNH 5088-97

The evolution of fish

[VERTEBRATE ANIMALS]

Fish were the first vertebrate animals and the ancestors of amphibians, reptiles, birds, and mammals. Many early fish had heads that were armored with bony plates.

One of the most fascinating aspects of the history of reefs is that although they have existed since the first algal mounds of the early Precambrian, subsequent reefs were constructed by an array of different plants and animals. Reefs formed by different organisms appeared in the early Paleozoic tropical seas. The reefs grew more complicated as reef organisms encrusted and overtopped each other. This process caused them to grow above the seafloor, providing platforms for animals to feed at different levels. Sponges, algae, and bryozoans were the main reef-builders in the Ordovician. Although we think of algae as seaweeds and soft, feathery scum that would not be capable of building anything sturdy, algae actually represent a much greater diversity of life forms. Some of the Paleozoic coralline algae were able to secrete calcium carbonate and create a hard rock-like structure that encrusted the reefs and bound them together.

The first fish

Fish first appeared in the seas of the early Ordovician period. Fish are the evolutionary ancestors of all other living vertebrates (animals with backbones): amphibians, reptiles, birds, and mammals. Today, they are the dominant predators in both fresh- and saltwater ecosystems. The earliest fossil fish are known from tiny bone fragments preserved in 500-million-year-old rock from Wyoming. Much better preserved fossils of some of the earliest known fish are found in a geologic formation known as the Stairway Sandstone, a section of which is exposed south of Alice Springs in the red center of Australia. These 480-million-year-old fish are preserved as sandstone casts of body segments. The Harding Sandstone near Cañon City, Colorado, also has preserved bones from some of these early fish. The first fish had external skeletons but no movable jaws. They lived near the muddy seafloor, sucking in organic debris through their jawless mouths. From this unassuming vacuum cleaner–like animal descended the jawed fish that became increasingly important in the waters of the world. There are over 21,000 species of fish today, making them the most diverse group of backboned animals alive.

High sea level and an ice age

The Ordovician was a time of extremely high sea level. Most of the continents were flooded. Thick deposits of Ordovician marine mudstone exposed in the deserts of Nevada and along the flanks of many ridges in the Appalachian Mountains are the evidence of great inundation. Because the continents were flooded, huge expanses of shallow marine settings provided a great variety of habitats. These extensive habitats allowed marine life to proliferate and evolve. Great amounts of space and the resulting abundance of resource allow for diversification; conversely, restricted space can contribute to extinction.

The mass extinction at the end of the Ordovician period snuffed out 22 percent of the families of marine organisms. Trilobites and brachiopods were decimated, and the marine ecosystems were greatly reduced in diversity. It is always difficult to determine the cause of a mass extinction, but there are several lines of evidence that point to falling sea level as the primary culprit in this case. We can see evidence of a period of extensive glaciation, an Ordovician ice age, in rocks of the Southern Hemisphere. At times of major glaciation, much of the world's water is trapped in the huge ice sheets that rest on the continents, and the sea level lowers accordingly. Near the end of the most recent ice age (about 12,000 years ago), the sea level was about 330 feet lower than it is today because so much water was trapped in the ice sheets that covered the northern parts of North America and Eurasia. If the sea level drops below the edges of the continents, the shallow marine habitat is greatly reduced. This is probably what happened at the end of the Ordovician. Animals that had adapted to life in the broad, shallow marine environments that covered most of what is now dry land suddenly found themselves with greatly shrunken habitats. This extinction was apparently caused by a space crisis!

After this ice age, animals that survived the extinction at the end of the Ordovician repopulated the world in a new way. The few lineages of brachiopods, trilobites, and cephalopods that survived into the Silurian period soon diversified into a whole new suite of related animals. In the tropical waters, reef ecosystems began to reassemble, but they were

Bony-headed fish

[Bothriolepis canadensis]

The bony head plates of early fish are more common as fossils than are complete fish. The fish probably had a soft body that rapidly decayed and fell off the sturdy armored head.

Late Devonian, 365 mya
Escuminac Formation, Canada
Fish length: 4.7 inches, DMNH 13192

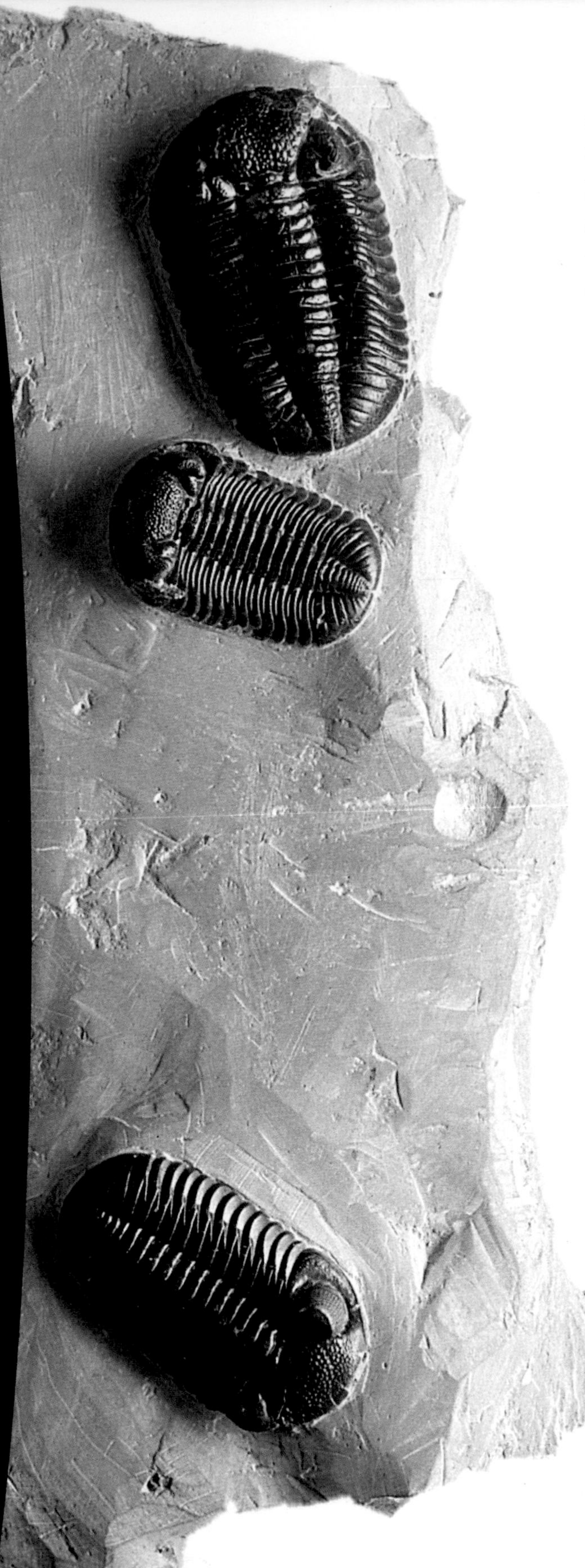

Trilobites

This slab of trilobites from Ohio is a graphic reminder that much of North America lay beneath shallow seas.

composed of different organisms than had populated the Ordovician reefs. Corals, sponge-like stromatoporoids, coralline algae, and crinoids built reef structures that had a passing similarity to modern reefs. The first fish with internal skeletons and functional jaws evolved and began to elbow their way into the shallow marine and freshwater environments. Cephalopods diversified into a myriad of shapes. Some of the straight-shelled cephalopods grew to lengths of 10 feet; their shells were the size of cannon barrels.

Trilobites remained the most common arthropods, but they were joined by chelicerates, the group that gave rise to today's horseshoe crabs, spiders, and scorpions, and by tiny bivalved crustaceans known as ostracods. Probably the best known Silurian animals are the eurypterids. Looking like a cross between scorpions and lobsters, eurypterids became major predators in shallow marine, brackish, and freshwater ecosystems. Pterygotid eurypterids grew as long as 5 feet, the size of a small person, and were equipped with huge lobster-like claws. Aquatic scorpions shared the same waters. It is interesting to find marine fossils of animals that today are totally terrestrial; such discoveries are one of the many lines of evidence that all land organisms have a marine heritage.

The Devonian period, which began about 410 million years ago, has been called the age of the fish because Devonian marine and freshwater deposits contain a rich array of fish fossils. Sharks made their first appearance and initial diversification in the Devonian. Another group of predatory fish, the placoderms, were armed with blade-like jawbones that functioned like huge teeth. The city parks of Cleveland, Ohio, contain cliffs of black Devonian marine shale that occasionally yield fossils of a huge placoderm fish known as *Dunkleosteus*. This massive beast was the first true giant of the seas. The largest *Dunkleosteus* is estimated to have been about 50 feet long—longer than a stretch limousine.

The arms race

The diversification of marine predators, including the cephalopods, eurypterids, and fish, made the oceans a more dangerous place and drove many organisms to develop defensive techniques. As predators became faster, larger, and stronger,

Trilobites

Early Devonian, 390 mya
Silica Shale, Ohio
Slab length: 13 inches,
DMNH 6011

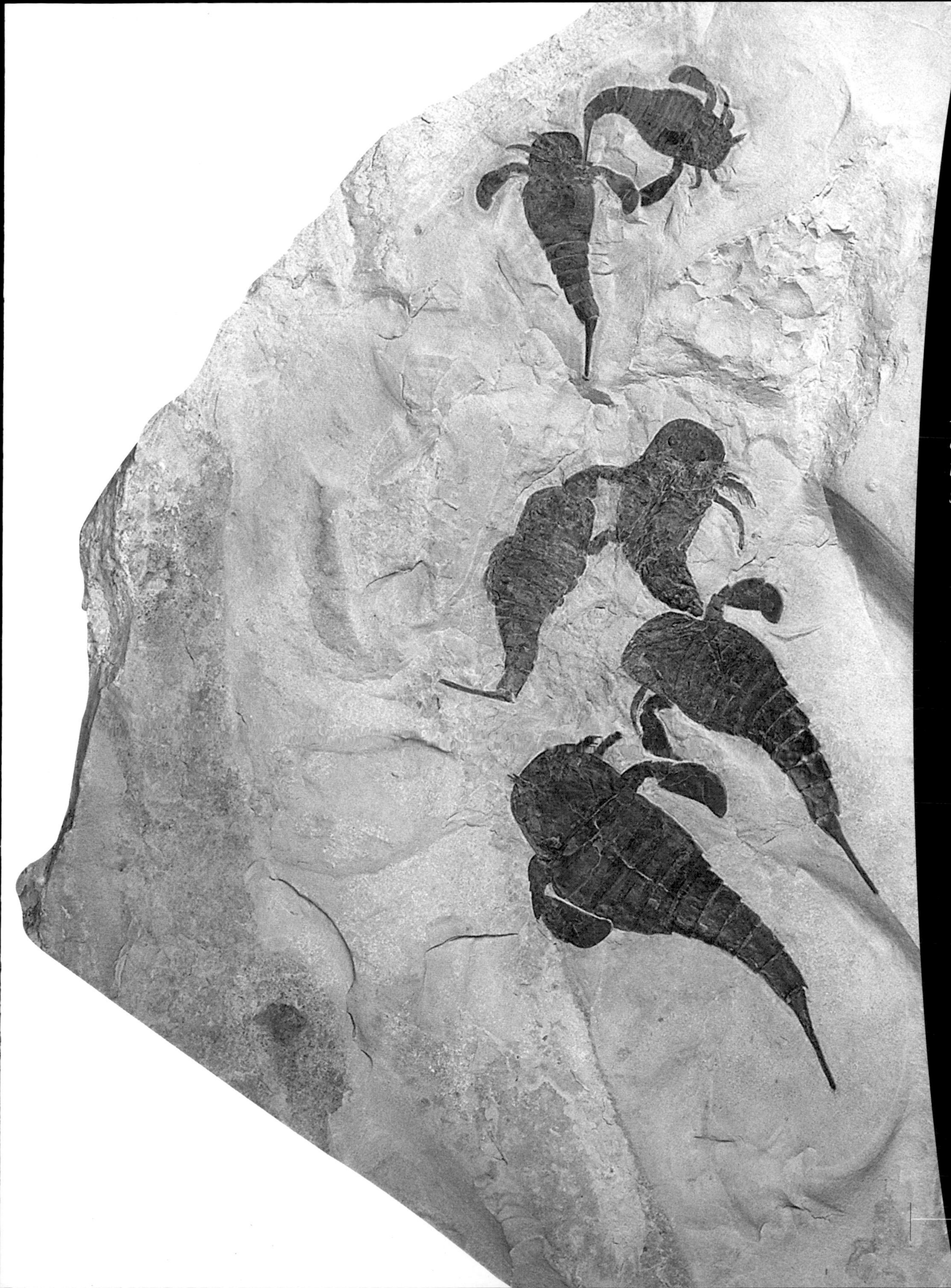

Sea scorpions

Eurypterids, or sea scorpions, swarmed the muddy bottoms of Silurian seas. Some were 5 feet long.

Sea scorpions

[*Eurypterus remipes*]

Early Silurian, 425 mya
Fiddler's Green Formation, New York
Length of longest animal: 11 inches,
DMNH 6001

their prey were forced to adapt or be eaten. Paleontologist Geerat Vermeij has likened the competition between predator and prey to the arms race and has suggested that it was the escalation of this race that led to the development of new offensive and defensive strategies and new body types. Some mechanisms, like defensive behavior, the ability to rapidly retreat, and cryptic coloring, are not commonly fossilized. However, there are defensive features, such as thick shells, spines, and burrows, that do fossilize. Fossils from the Devonian period show an array of these features that can be attributed to the arms race between predator and prey.

Trilobites, everyone's favorite fossil of the Paleozoic, exhibit a variety of body types and behaviors that suggest they were responding to aggressive predation. Many Paleozoic trilobites had the ability to roll up their bodies like modern armadillos and pill bugs. This behavior would protect the animal's soft underbelly and is often fossilized in the form of enrolled trilobites. The spines of *Dicranurus,* a Devonian trilobite, were twice the length of its body. An enrolled *Dicranurus* would present a potential predator with the marine equivalent of a porcupine. In less defensive moments, the spines may have functioned like snowshoes to keep the trilobite from sinking into the soft mud of the seafloor. Trilobites such as *Arctinurus* often are preserved with bite-marks on their shells, suggesting that it was not uncommon for them to survive attacks. Other trilobites may have avoided attack by hiding. *Neoasaphus,* an Ordovician trilobite, carried its eyes at the end of stalks that were nearly the length of its body, allowing the animal to lie beneath the mud of the seafloor and still survey its surroundings.

Like trilobites, brachiopods, a group of marine animals that may be the most common and diverse of the fossils found in Paleozoic rocks, evolved a variety of body shapes in response to predation and the environment. Some developed thick shells while others had long spines. Brachiopods look superficially like clams but are radically different animals. Although they have paired shells, each side is not the mirror image of the other. Brachiopod fossils are so diverse and widespread that they are useful for dating the rocks in which they are found. Toward the end of the Paleozoic, some

Stalk-eyed trilobite

[Neoasaphus kowalewskii]

The eyes of this trilobite are mounted at the ends of inch-long stalks. Thus, the animal could conceal itself in the muddy seafloor and still observe its surroundings.

Middle Ordovician, 465 mya
St. Petersburg, Russia
Trilobite length: 3.2 inches, DMNH 7813

Spiny trilobite

[Dicranurus hamatus]

The long spines of this trilobite more than double its body length. They may have served as defense or as "snowshoes" for traveling on a soft and muddy seafloor.

Early Devonian, 390 mya
Haragan Formation, Oklahoma
Trilobite length: 3.7 inches, DMNH 6251

brachiopods became major elements in the construction of reefs. These reef-building brachiopods had bizarre elongated shells that were reminiscent of the cone-shaped, reef-building archaeocyathans of the early Cambrian and the similarly shaped rugose corals of the Silurian and Devonian periods.

Looking toward land

By the middle of the Silurian period, life was a bustling business in the seas and oceans of the world, but the land masses remained strangely unpopulated. Crowded nearshore ecosystems lived within sight of a great deal of vacant real estate on land. But the challenges of life on land were many. The seawater sheltered organisms from the sun's deadly ultraviolet radiation and provided a continuous and surrounding source of moisture and nutrients. The seawater was the medium in which life had evolved and diversified. But the environment at the edge of the sea is and was very diverse, ranging from sandy beaches to rocky shorelines, tidal flats, deltas, estuaries, and coastal bays. As organisms adapted to life in these marginal habitats, they were unwittingly preparing themselves for life on land. On tidal flats they were occasionally exposed to the air when tides were low, in the coastal estuaries they were exposed to fluctuating salinities, and on the rocky shores they were battered by crashing waves. The Paleozoic was a time of fluctuating sea levels, and sometime during the Silurian period, when the sea level fell and the coastline moved seaward, plants and animals began to live on land.

Spiny trilobite

[Psychopyge elegans]

The spikes and spines of this elegant trilobite were a challenge to the fossil preparator, who spent dozens of hours using a needle to free this animal from its rocky tomb.

Early Devonian, 390 mya
Hamar Laghdad Formation, Morocco
Trilobite length: 4.3 inches, DMNH 6172

Beartooth Butte, Wyoming

[395 Million Years Ago]

The land is nearly barren. Bright white limestone cliffs surround the estuary that leads to the ocean. There is virtually no hint of life on land. No buzz of insects, no calls of birds, no rustling of leaves—only the rhythmic breach of waves against a rocky, muddy shore. At the water's edge, a network of primitive green plants breaks the monotony of earth colors. Three types of plants, Drepanophycus, Psilophyton, *and* Rebuchia, *lace the edge of the water where it meets the cliffs. Green tendrils reach upward; some are topped with round clumps of life-bearing spores called sporangia. Crawling among the plants are barely visible, insect-like arthropods and much more conspicuous scorpions. Many of the plants have been flattened by a recent flood. Some are coated with mud while others float atop the water. The jawless fish* Protaspis *and the lungfish* Uranolophus *abound in the murky water of the estuary. Bony head plates and carcasses of some of their dead companions lie in the debris washed onto shore. An aquatic lobster-like eurypterid picks at a submerged fish carcass. The first invertebrates and plants have approached land, and soon they will fill it.*

chapter FOUR 4

Between Two Worlds:

The First Life on Land

Perched high on Wyoming's Beartooth Plateau, nearly 12,000 feet above sea level and in the midst of prime grizzly bear habitat, Beartooth Butte seems like an odd place to look for evidence of the first life on land. In order to reach the site, you have to trek across a lush alpine meadow and scramble up an unstable talus slope composed of teetering blocks of limestone and dolomite. Once you reach the base of the cliff, the only way up the face is through an active debris chute down which large boulders regularly rumble. If you brave this danger and scramble up onto the red outcrop, you will be rewarded with a spectacular view of the jagged Absaroka Range to the south. At your feet are red rocks that were once estuarine mud on a sunbaked shore where the first forms of life braved the elements and attempted to exist on land.

Beartooth Butte is one of only a handful of sites that provide evidence for some of the earliest forms of terrestrial life. Rocks of the Devonian Beartooth Butte Formation formed by the cementing of sediment that had accumulated in the channels of estuaries that flowed into the Devonian seas. The fossils at Beartooth Butte were discovered in 1931 by Erling Dorf, a paleontologist from Princeton University. The plant fossils he found were considered at that time to be the world's oldest terrestrial vegetation. Since then, new findings from other sites and our ability to identify the microscopic spores of early land plants have pushed back evidence of the first colonization of land well into the Ordovician period.

The fossil plants of Beartooth Butte are preserved as carbonized compressions of partial stems. They were, when alive, green plants that stood a few inches to a few feet tall. The plants *Psilophyton* and *Rebuchia* are very simple in structure: slim green stems with no leaves, roots, or seeds. The more complicated *Drepanophycus*, a primitive club moss, or lycopod, sports small leaves and the world's first roots. The most common animal fossils are the scattered head and body plates of jawless fishes. Arthropods are represented by scorpion and eurypterid bodies and parts. When dissolved in hydrofluoric acid, the mudstone matrix yields fossil fragments of minute mite-like arthropods, some of the first animals on land.

Eurypterid claw

[Pterygotid eurypterid]

This claw belonged to a dangerous predator that was almost 5 feet long.

Early Devonian, 395 mya
Beartooth Butte Formation, Wyoming
Claw length: 7.3 inches, DMNH 6357

Primitive plant

[*Psilophyton wyomingensis*]

Although it had no leaves, roots, or seeds, this early plant was able to colonize land.

Early Devonian, 395 mya
Beartooth Butte Formation, Wyoming
Slab length: 5.9 inches, DMNH 6381

Fish head plates

[*Protaspis transversa*]

The earliest fishes had flattened bodies and heads armored by bone plates.

Early Devonian, 395 mya
Beartooth Butte Formation, Wyoming
Slab length: 18.7 inches, DMNH 8787

Lycopod stem

[*Drepanophycus devonicus*]

The spiky projections on this early club moss stem are some of the world's first leaves.

Early Devonian, 395 mya
Beartooth Butte Formation, Wyoming
Stem length: 2.2 inches, DMNH 6365

chapter FOUR

Between Two Worlds:

The First Life on Land

The Devonian World

[395 MILLION YEARS AGO]

Continents begin to drift together and shallow seas cover portions of the land masses. Plants, insects, and other arthropods colonize the land.

Adaptation to life on land

Organisms living in water have an easy life. The water supports them, bathes them in moisture, provides nutrients, and washes away their waste products. The water even shields them from the sun's harsh ultraviolet light. In order to live on land, plants and animals must abandon all of these maritime perks. Organisms must evolve and adapt to secure and transport water and nutrients, to dispose of waste, to prevent drying out, and to support body weight.

Evidence of the first life on land has been revealed through an assortment of minute clues that can be seen only through the microscope. Ancient soil deposits and the burrows of what appear to have been terrestrial organisms are known from rocks older than the first actual remains of either terrestrial plants or animals. Like the first life forms in the oceans, some of the earliest forms of terrestrial life were probably mats of microbial life—tiny cyanobacteria. Perhaps as early as the Precambrian, these mats may have existed in protected habitats near water where the effects of extremes in climate and physical devastation would have been less severe. The cell structure and biochemistry of green algae are quite similar to those of terrestrial plants, and at some point during the Ordovician or Silurian period, green algae gave rise to the first vascular land plants. The first known remains of land plants are microscopic spores and tiny sheets of waxy stem and leaf coating, or cuticle, that remained after softer body parts had decayed.

The first complete vascular plants did not appear until about 420 million years ago in the late Silurian. These plants had cuticle, which prevented drying out, and a vascular system for transporting water and nutrients. The vascular system had rigidity that allowed the small plants to stand erect, which enabled them to capture more sunlight. One of the earliest vascular plants was an unassuming 3-inch-tall herb known as *Cooksonia.* The forked and leafless stems of this plant were photosynthetic and were topped by rounded structures that produced microscopic spores. The tiny spores were somewhat resistant to drying and traveled well in the wind, allowing them to spread across the landscape. It was not long before a moderate diversity of small plants appeared, forming the first simple terrestrial ecosystem.

The Devonian period was to land plants what the Cambrian period was to marine invertebrates: a time of rapid evolution and innovation. During the Devonian period, many of the characteristics of modern plants first appeared—leaves, roots, seeds, and wood. With improved water-transporting vessels and sturdy wood, plants competed for light and became taller. By the end of the period, the first forests covered the landscape, providing the first shade. Sometime near the end of the Devonian, some plants developed seeds. The hard outer coating of seeds prevented them from drying out, protected the plant embryo, and preserved a small storehouse of food. Seeds allowed plants to grow in shade or where moisture was variable. Now the plants had a method of reproduction so they truly could inhabit the land. By the end of the Devonian, landscapes were filled with a diversity of tree-sized ferns, horsetail rushes, scaly lycopods, and progymnosperms, trees with tiny fan-shaped leaves and the first true wood.

Arthropods were the first terrestrial animals. Fragments of their tiny external skeletons, or exoskeletons, have been discovered in rocks of early Silurian age in Pennsylvania. The exoskeletons of the early arthropods served as excellent protection against ultraviolet radiation as well as providing a rigid body to cope with the force of gravity. Bits of arthropod exoskeleton are common fossils in Silurian rocks. Etched from the rock with acids, the pieces look like segments of modern insects. More complete fossils of centipedes are known from 415-million-year-old rocks in

Animals
lonize land

THE FIRST
PHIBIANS]

ibians were the first
rate animals to live
nd, but their repro-
ve cycle kept them
t on standing bodies
ter. When they first
red onto land, am-
ans entered bustling
stems already alive
nsects and arachnids.

Progymnosperm leaves

[Archaeopteris macilenta]

These fern-like leaves seem unsuited to the plant that bore them: a giant tree with conifer-like wood.

Late Devonian, 360 mya
Pocono Sandstone, Pennsylvania
Slab length: 13.3 inches, DMNH 5849

England. Spider-like creatures called trigonotarbids, which have been discovered in Scotland and New York, date from the early Devonian, or about 380 million to 400 million years ago. These early insects and arachnids preyed on each other beneath the lilliputian canopy of 1-foot-tall plants.

Early plants provided a good source of nutrients for a variety of terrestrial arthropods, including millipedes, centipedes, and insects. These creatures did not, however, directly eat plants but rather fed on the fallen leaves and stems that decayed as detritus on the ground. Evidence of an animal feeding on a live plant does not occur until the Pennsylvanian period, 50 million years later.

Origin of terrestrial vertebrates

The earliest vertebrates to inhabit land, the amphibians, appear much later in the fossil record than the first land plants and invertebrates. Their fossils are found in rocks from the late Devonian, about 375 million years ago. These early tetrapods, vertebrates with four legs, possessed features that enabled them to live on land, including adaptations to move, feed, and breathe outside of water.

The earliest tetrapods resembled modern salamanders. They possessed a strongly supported skull; fully developed front and hind limbs with up to eight digits, rather than five, on each foot; a rather flat body; and a tadpole-like tail. Although there was much debate during the 1970s and 1980s about which group was ancestral among the fishes to the first tetrapod, paleontologists now agree that it was a group known as the lobe-finned fish, which today are represented by the coelocanth.

Among the lobe-finned fishes, one group in particular, the osteolepiforms, is closely related to and may be ancestral stock for the first tetrapods. It is quite remarkable how closely the osteolepiforms and earliest amphibians resemble each other. Although osteolepiforms are clearly fish with fins and gills and the earliest tetrapods are amphibians with four limbs and a rigid skull, they share many characteristics. The skulls of both are nearly identical and have the same arrangement of bones. In some cases the bones of the skulls of both osteolepiforms and early tetrapods are unique and do not appear in any other group of fish. As the name *lobe-finned* implies, the fins of the osteolepiforms were rather stout and limb-like. The foreleg of the amphibian and fore-fin of the osteolepiform both bear one single upper arm bone and two lower arm bones. These are equivalent to the humerus, radius, and ulna of all modern tetrapods. The vertebrae of both creatures bear similar arches, and parts of the vertebral body are very similar in form. The osteolepiforms have a rich fossil record during the late Devonian and throughout the Mississippian and Pennsylvanian, overlaping that of the early tetrapods.

The earliest tetrapods have advanced characteristics that are shared by all amphibians, reptiles, mammals, and birds, which demonstrates that all of them descended from a common ancestor. The adult amphibian skull is rigid and the gill mechanism has been lost and replaced by nostrils and lungs. But a notch remains in the early tetrapod skull where the gill mechanism was present in fish. The limbs are clearly fins in osteolepiforms with many bony rays, whereas in the earliest tetrapods, up to eight digits take the place of the rays. Both the front and the hind limbs of the earliest tetrapods were well built for moving the animal across a hard surface. The shoulder girdle was free from the skull rather than attached to it, and the bones of the hip were rigidly attached to the vertebral column. The bodies of the earliest amphibians were relatively low and flat with stout ribs to aid in breathing air. By looking at living amphibians such as frogs and salamanders, we can speculate that the earliest tetrapods probably reproduced in a manner like that of many fish; the abundant eggs were fertilized and developed to the adult stage in water.

Many of the hallmark features of tetrapods may have evolved prior to their full adaptation to land. The tetrapod limb would have functioned very well to move the animal around in shallow pools of water and may actually have been more efficient than the fins of fish in this situation. Similarly, breathing air would have given them a strong selective advantage in habitats where the water was shallow or poorly oxygenated. Why then did the vertebrates move onto land? Perhaps it was because the adaptations that allowed them to live in shallow, poorly oxygenated pools were much the same as the ones that they would use to exploit the untouched food resources on land.

By the end of the Paleozoic, many different amphibian groups had evolved, some with small snake-like bodies, others with boomerang-shaped heads, and others of large proportions similar to alligators. With adaptations for terrestrial life, the stage was set for the development of the amniotes—the group of tetrapods that could lay eggs on land and thus be free from having to live near a relatively permanent water source. The earliest amniotes appeared in the Carboniferous period. They were the common ancestor of the more advanced terrestrial vertebrates that today include reptiles, birds, and mammals.

Like the arthropod colonists of the land, the early terrestrial vertebrates appear to have fed on insects and other terrestrial arthropods, but they also fed on fish and perhaps each other. Plants did provide the basis for the terrestrial food chain but not directly. Many of the early terrestrial arthropods were detritivores, consuming plant matter that was partially decayed and rotted. The early forested landscapes would provide numerous opportunities for new lifestyles to evolve and new ecosystems to emerge.

Lobe-finned fish

[Eusthenopteron foordi]

The lobe-finned fishes were diverse in the Paleozoic but today are represented by only a single species. They were the ancestors of all four-footed beasts, including the amphibians, reptiles, birds, and mammals.

Late Devonian, 365 mya
Escuminac Formation, Quebec, Canada
Fish length: 10.2 inches, DMNH 2242

amilton,
Kansas

[295 Million Years Ago]

A small coastal stream winds hrough a narrow valley fringed y limestone ledges. The sun is set-ing and a calm sea is visible in he distance. Forest covers the andscape, but the plants and ani-nals in it are not familiar to the nodern eye. The streambed sup-orts a forest of lycopods, trees that ook like scaly telephone poles, and teridosperms, plants that look ike tree ferns except for their con-spicuous seeds. Amphibians lounge ear pools of water in the drying treambed while small but danger-ous fin-backed pelycosaurs prowl he limestone outcrop and the for-est floor. The scene is alive with the ounds of insects, rustling trees, and the far-off rhythm of surf hit-ing the beach. Crickets are heard as is the occasional whir of giant dragonflies in flight. Cockroaches are abundant, some reaching 3 inches in length, and 2-foot-long nillipedes wander through the eaf litter. Here, the insects are early as large as the vertebrate animals. The forest on the drier imestone ledges is dominated by eed-cone-bearing cordaites and Walchia, the earliest conifer. A ire has ravaged parts of the for-st, leaving many of the trees charred and defoliated.

chapter FIVE

Forests and Flight: Life in the Late Paleozoic

The Flint Hills area of eastern Kansas is a landscape of wooded farmland and Norman Rockwell towns. In this classic small-town setting, tractors are parked in front of cafes and farmers still raise cows, corn, and more farmers. Just beneath this pastoral veneer, near the tiny town of Hamilton, lie the remains of a 295-million-year-old coastal stream. In general, the rocks in this part of Kansas, layers of marine limestone and shale, are full of fossils of animals that lived in the Pennsylvanian seas. Most outcrops yield brachiopods and crinoid stems. But the Hamilton site is different. Here the rocks are of three kinds: a massive limestone full of marine fossils, a blocky conglomerate made up of chunks of this limestone, and a tightly laminated limestone with peculiarly regular layering. The key to understanding the Hamilton site is the geometry of these rock layers. The conglomerate lies in a layer below the laminated limestone in an elongate, lens-shaped deposit that is inset into the massive limestone. This relationship has been enhanced by road crews that quarried out the conglomerate, leaving a long shallow valley. Unwittingly, these quarry crews have cleaned the debris out of an ancient stream valley. Careful examination of the limestone that lies on top of the conglomerate shows that it was deposited by regular tidal cycles. Each lamination represents the deposit of a single tide, and these layers were deposited in sets that correspond to the spring and neap of a monthly tidal cycle.

The tidal layers are like little clocks, their organization indicating the speed at which they had been deposited. Because rapid deposition is one very successful way to preserve fossils, these lay ers have more than their share. Preserved in the tidal deposits are the bones and skeletons of fish, sharks, amphibians, reptiles, fin-backed protomammals, and the bodies, wings, and carapaces (the upper shells) of scorpions, eurypterids, millipedes, dragonflies, crickets, and cockroaches. Also common are the strap-like leaves of the cordaites and the scaly branches and cones of *Walchia.* Less common are the radially symmetrical leaves of sphenopsids, the leaves and seeds of seed ferns, and slabs of bark and root pieces from lycopod trees. Among these fossils are briquette-like chunks of fossil charcoal, evidence of seasonal dry periods and resulting forest fires.

Conifer branch

[Walchia piniformis]

This battered frond fell from one of the first conifers.

Late Pennsylvanian, 295 mya
Topeka Limestone, Kansas
Branch length: 9.8 inches, DMNH 7024

Cordaite leaf

[Cordaites principalis]

With large strap-like leaves and seed-bearing cones, cordaites were among the most common seed plants in the Pennsylvanian swamps.

Late Pennsylvanian, 295 mya
Topeka Limestone, Kansas
Leaf length: 2.1 feet, DMNH 5100

Cockroach

[Orthomylacrid cockroach]

Then, as now, cockroaches were common in tropical settings.

Late Pennsylvanian, 295 mya
Topeka Limestone, Kansas
Insect length: 2.9 inches, DMNH 7023

Protomammal jaw

[Ophiacodont pelycosaur]

The mid-Paleozoic land communities had many predators that lived on detritus-feeding insects and invertebrates. The sharp, blade-like teeth of this protomammal testifies to its predaceous habit.

Late Pennsylvanian, 295 mya
Topeka Limestone, Kansas
Jaw length: 4.3 inches, DMNH 12844

5

chapter FIVE

Forests and Flight:

Life in the Late Paleozoic

The Carboniferous World

[295 MILLION YEARS AGO]

Ice on the South Pole creates glaciers and affects global climate as the continents drift together to form the supercontinent known as Pangaea.

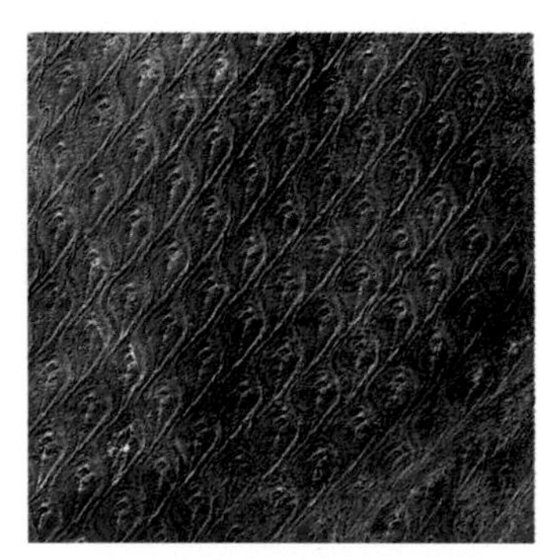

Surface of a lycopod tree trunk

[*Lepidodendron modulatum*]

The scales of the lycopod tree bark are actually scars from where grass-like leaves have fallen.

Late Pennsylvanian, 300 mya
Llewellyn Formation, Pennsylvania
Average scale length: 1.2 inches, DMNH 929

The time of swamps

At the time that the Hamilton ecosystem existed, extensive coal-forming swamps were present in what would become the eastern United States and western Europe. These sites, too, were coastal, but they were also equatorial and apparently were not as affected by the seasonal dryness that characterized the Hamilton site.

The Carboniferous, which is divided into the Mississippian and Pennsylvanian periods in North America, began about 360 million years ago. It is, as its name suggests, famous for the huge and extensive coal deposits that formed during this time in Pennsylvania, Illinois, Kentucky, West Virginia, Nova Scotia, England, France, Germany, Poland, and Russia. Many people call this period the Coal Age.

In the Carboniferous, forests looked completely different than they do today because the trees were so unlike the ones we know. Our forests are made of either conifers (such as pine, spruce, and hemlock) or broad-leafed flowering plants (such as oak, maple, and aspen). In those early forests, the dominant trees were lycopods, calamites, ferns, pteridosperms, and cordaites. Calamites, lycopods, and ferns are plants that reproduce with spores and whose living relatives are generally herbaceous ground plants with the notable exception of the tree ferns. Pteridosperms are often called seed ferns because they have fern-like foliage and, surprisingly, seeds. The cordaites were seed-bearing plants with strap-like leaves and conifer-like wood that became extinct before the end of the Triassic period. Only tree ferns have survived as a visual reminder of the Carboniferous coal forests, and modern tree ferns are quite distant relatives of their Carboniferous cousins.

The lycopods were perhaps the strangest of all the Carboniferous plants. Lycopod trees were an amalgam of odd and distinctive parts, perhaps the most distinctive of which is the patterned bark that looks more like reptilian scaly skin than the surface of a tree trunk. The scales are actually leaf bases from which long whip-like leaves once emerged. Lycopods had large cones that bore spores rather than seeds. The base of the lycopod was a bifurcating underground network known as the stigmarian system. The trunk of the tree was composed of a surprisingly thick layer of bark that took the place of wood as the main mechanism for holding up the tree.

Some lycopods, like *Lepidodendron,* had a trunk that forked repeatedly to create a skeletal canopy. Others, like *Sigillaria,* formed tall straight poles that forked only once or not at all. Lycopods had very little of the plant tissue that transports the products of photosynthesis in modern plants, suggesting that many parts of lycopod trees were photosynthetic and were able to make food. Thus, most of the lycopod tree—even some of its roots—would have been green! It seems that these trees also grew backward. A modern tree such as a conifer grows up and out at more or less the same time. Lycopods seem to have developed their stigmarian root system first. Then the trunk shot up at its maximum diameter by adding plant tissue at the top. Forests of lycopods, called pole-tree forests, would have been strange sights indeed. Imagine a vast area populated by giant 100-foot pipe cleaners, and you have an image of the pole-tree swamp forests of the Carboniferous.

In a pole-tree forest, sunlight would not have been impeded by a canopy and would have reached the forest floor at full strength. A tree that is photosynthetic from its leaves all the way down to its roots seems like a much more viable proposition in such a situation than it would be in the darkness of a modern forest floor. A grove of these trees rooted in the soft peaty soil of a swamp would be susceptible to blowdowns in windstorms. A strip mine in northern Pennsylvania preserves the imprint of a whole

Trees cover the land

[CARBONIFEROUS FORESTS]

Three hundred and fifty million years ago, the world was covered by forests for the first time. The trees that made up these forests included lycopods (scale trees), giant horsetails, ferns, fern-like seed plants known as pteridosperms, cordaites, and the first conifers.

Forest floor leaf litter

This leaf litter from an ancient Pennsylvanian swamp was replaced by a white clay during fossilization. These fern-like leaves grew on extinct seed plants known as pteridosperms.

Forest floor leaf litter

[*Alethopteris serlii* and *Trigonocarpus* sp.]

Late Pennsylvanian, 300 mya
Llewellyn Formation, Pennsylvania
Average frond width: 2 inches,
DMNH 6809

lycopod forest that was knocked down by such a storm. The alignment of the fossil trees in this forest made it possible to ascertain the travel direction of the storm that blew up more than 300 million years ago!

Amphibians and reptiles

The animals that lived in these pole-tree forests included insects and other arthropods, amphibians, and the first reptiles. Reptiles, birds, and mammals form the single group Amniota, which is distinguished by a method of reproduction independent of standing water. The young are born or hatched as miniatures of the adult rather than having a tadpole larval stage. Reptiles were the first animals to produce an amniote egg, which was a more durable package than the amphibian egg. We do not know when the first reptile egg was laid, but the first reptile fossils appeared in the late Mississippian and belonged to a group known as the captorhinomorphs—small animals with unspecialized dentition that probably fed on insects. By the late Carboniferous, the primitive captorhinomorphs had evolved into three groups: one that led to mammals, one that led to turtles, and one that led to all other reptile groups and birds.

Fossils of some of the earliest reptiles have been found in the Pennsylvanian coalfields near Joggins, Nova Scotia. The small skeletons are found in the stumps of giant lycopods. Apparently, the upright trees were buried in mud as they stood, so when they died and rotted away, they left a hollow vertical shaft, which trapped reptiles that were unlucky enough to fall into the hole. Later, floods filled the hollow and encased the reptile skeletons in a mud cast of the tree trunk.

Reptiles and their descendants have three basic skull types: the anapsid type, which includes the earliest reptiles and modern turtles; the synapsid type, which includes protomammals (formerly called mammal-like reptiles) and mammals; and the diapsid type, which includes all living reptiles other than turtles and the extinct groups of reptiles

Boomerang-head amphibian

Amphibians were the first four-footed beasts to live on land, but they were equally at home in the water. The boomerang-head amphibian spent more time in the water than on land.

Boomerang-head amphibian

[Diplocaulus sp.]

Early Permian, 275 mya
Vale Formation, Texas
Length: 2.7 feet, DMNH 8905/6

Showdown in old Texas

[DIMETRODON AND ERYOPS]

The protomammal Dimetrodon *springs from a hiding place in the gigantopterid foliage and attacks the big-headed amphibian* Eryops.

Fin-backed protomammal and big-headed amphibian

[*Dimetrodon limbatus* and *Eryops megacephalus*]

Dimetrodon, a protomammal, and *Eryops*, an amphibian, are common fossils in the Permian redbeds of Texas and Oklahoma.

Early Permian, 275 mya
Admiral Formation, Texas
Length: 6.2 feet, DMNH 10803
Length: 6 feet, DMNH 10804

known as ichthyosaurs, nothosaurs, and plesiosaurs. The synapsids consist of three groups: the pelycosaurs, the therapsids, and the mammals. The pelycosaurs are known from the same time period (Pennsylvanian) as the earliest conventional reptiles. Both the pelycosaurs and the therapsids included a wide array of adaptive types including omnivorous, carnivorous, and insectivorous groups.

The teeth of the amphibians and reptiles found in Joggins suggest that they ate insects or each other. There does not appear to be any evidence for plant-eating amphibians or reptiles in the Carboniferous, a situation that differs markedly from the modern world. Thus, the food value of plants could not pass directly into the animals; it had to travel via an indirect route, probably through detritus-eating insects. Although there is little evidence for insect herbivory on living plant tissue at this time, it does seem likely that insects consumed forest litter after it had been partially decayed by bacteria. Some reproductive parts of plants, such as the spore sacs, appear to have been attacked by insects.

Giant spider

[Megarachne servinei]

Arachnophobia is guaranteed in the presence of this giant, 14-inch-long, tarantula-like spider found in Argentina. Swamps of the late Paleozoic were creeping with giant insects and arachnids.

Pennsylvanian, 310 mya
Bajo de Veliz Formation, Argentina
Spider length: 13.4 inches, DMNH 7800

Giant insects

A surprising and appalling aspect of the insects and other terrestrial arthropods of the Carboniferous is their size. Coal miners in France in the late 1800s uncovered some of the first fossil insect giants: a dragonfly-like creature that had a 2-foot wingspan. Carboniferous rocks in Argentina have yielded a tarantula-like spider with a 14-inch body! From Nova Scotia come the remains of a millipede-like creature called an arthropleuran whose body was more than 6 feet long and 1 foot wide. And some of the early scorpions extended well past the 2-foot mark. What encouraged this insect gigantism? Why don't such giants exist today? These questions remain unanswered, but some clues might come from the rarity of vertebrate predators or a different composition of the earth's atmosphere that allowed more efficient breathing and growth among these insect giants.

Somehow, these insects began to fly, which helped them escape from predators, forage, and travel. The origin of flight is a mystery, but there are several theories. The most likely is that flight came about when insects ran, jumped, and were borne aloft by air currents or parachuted from high spots. Those insects with body parts that allowed for more efficient air travel survived and reproduced more often than their less endowed colleagues. Eventually the fine blade of evolution carved insects with the ability to power their own flight. The question of how flight originated has to be asked many times because flight in earth's history appeared independently in insects, reptiles, birds, mammals, and even fish.

Dry times on a supercontinent

If the amniote egg was the adaptation that provided the reptiles with the ability to colonize new areas, the seed did the same thing for land plants. Although the first seeds had appeared in the late Devonian, they really found their place in the Carboniferous landscape with the spread of pteridosperms, cordaites, and the new kids on the block: the conifers.

The development of amniote eggs in reptiles and of seeds in plants was timely. Toward the end of the Carboniferous, all of the world's continents were drifting together to form one huge supercontinent known as Pangaea. The formation of one continent had extreme climatic effects as the concept of continentality got its first big test. Seasonality and aridity became stresses that had to be contended with. Geologic evidence from all of the southern continents suggests north-flowing ice sheets. In fact, some of the first evidence of continental drift was the presence of late Carboniferous glacial deposits in Australia, South Africa, India, and Antarctica. Through a combination of decreasing carbon dioxide in the atmosphere and increasing continentality, the climate began to take a turn for the worse. The coals of the Carboniferous gave way to the arid deposits of the Permian.

The end of the Carboniferous period and the beginning of the Permian, about 290 million years ago, saw the demise of many of the moisture-loving plants. Gone were most of the lycopods and calamites. They were replaced by conifers and a new group of plants that would eventually give rise to the cycads. As the conditions became drier and more seasonal, seed plants had an advantage over the water-loving ferns, lycopods, and sphenopsids.

The protomammals

[AULACEPHALODON, THRINAXODON, AND DIMETRODON]

A diverse group of Permian four-legged animals, the protomammals, were the dominant land animals before the rise of the dinosaurs. The protomammals became extinct but they gave rise to the mammals that were to diversify after the extinction of the dinosaurs

Glass Mountain reef

This block of brachiopods, corals, and bryozoans from the Glass Mountains of Texas was soaked in acid to dissolve the limestone matrix from the delicate shells. The shells had naturally been replaced with silica, making them resistant to the corrosive acids. In the Permian period, these animals lived in a tropical reef.

Glass Mountain reef

[*Collemataria, Meekella, Grandaurispina* and *Paucispinifera*]

Early Permian, 275 mya
Gaptank Formation, Texas
Specimen length: 14.4 inches,
DMNH 5102

Extensive deposits of Permian rocks in Texas and Oklahoma have produced both plant and animal fossils. By the Permian, vertebrates had figured out how to eat plants and had thus done away with the insect middleman. Finally, the green planet was a salad. Some of the first herbivorous vertebrates were fin-backed edaphosaurs, related to the pelycosaur *Dimetrodon,* and the beak-nosed protomammals. Superb records of these new guilds of animals come from rocks in Russia, South Africa, and the American Southwest. The pelycosaurs diversified during the early Permian, as did the therapsids in the late Permian. During the Permian, protomammals dominated many faunas. They include a bizarre array of organisms that were as exotic and strange as their names: varanopsids, caseids, anomodonts, dinocephalians, venjukoviamorphs, gorgonopsians, and gomphodonts, among others.

The plants of the Permian represent adaptation to the new arid landscape. Conifers, which had their start in the late Carboniferous, became common on the landscape while lycopods and cordaites became rare. Pteridosperms survived and gave rise to new groups such as the feathery callipterids. The cycads and taeniopterids, gymnosperms that became increasingly important in the Mesozoic, got started in the Permian. Permian floras also contain a number of plants of questionable affinity (that is, we don't know what they are related to). These include the gigantopterids, plants that had large compound leaves or simple large bilobed leaves. These plants, found in China and North America, may have been large fleshy herbaceous plants, or they may have been vines.

The end-Permian extinction

Like all good things, the Paleozoic era came to a close, and it did so in spectacular fashion in the form of the end-Permian extinction. At no other time in earth's history has life come so close to getting exterminated. Fully 96 percent of marine invertebrate species did not survive the end-Permian extinction. Among the marine organisms, 98 percent of the crinoid families, 96 percent of tabulate and rugose corals, 78 percent of articulate brachiopods, 76 percent of bryozoans, 71 percent of cephalopods, 50 percent of foraminifera, and all trilobites became extinct. Hardest hit were the animals that lived attached to the seafloor, such as the crinoids, bryozoans, and corals. Some groups fared well, including gastropods, sponges, and bivalves. On land, floras of the Paleozoic were gradually replaced by those of the Mesozoic. Permian amphibians and therapsid protomammals were replaced by different therapsids and early diapsid reptiles such as the thecodonts, which were the ancestors of the dinosaurs.

The cause of the Permian extinction is, like the causes of most extinctions, hotly debated. The glaciation at the end of the Carboniferous had ceased by the middle of the Permian, so it has an alibi and probably was not the culprit. In northern Asia, a huge lava field, known as the Siberian traps, formed at the end of the Permian, and explosive volcanism shook southern China. The volcanic gases released by these events may well have had an unusual effect on the earth's atmosphere and caused dramatic and deadly climatic changes. But rapidly falling sea level, or regression, also occurred at this time. Regressions, by definition, reduce the amount of shallow-water marine habitat. With no place to live or go, many marine organisms perished. Associated with rapid sea-level fluctuations were phases of seawater anoxia, during which the ocean's oxygen content dropped to levels that were lethal. No one single cause of the extinction stands out; instead there appears to be "a complex web of causality," a phrase coined by Douglas Erwin, a paleontologist at the Smithsonian Institution. It's another way of saying that the greatest mass murder in the history of the earth remains an unsolved crime.

Petrified Forest National Park, Arizona

[225 Million Years Ago]

The huge and well-toothed iead of a large crocodile-like ɔhytosaur splashes water as he massive beast lunges from ɔeneath the oily surface of a ɔlack water slough. The shade ɔrovided by the overarching hickets of the giant horsetail Neocalamites *allowed the phyto-aur to make a stealthy approach ɔn a preoccupied pack of feed-ng* Coelophysis *dinosaurs. His ntended prey is alert and nimble ɛnough to spring out of harm's vay. The* Coelophysis *pack had ɛmerged from the riverside jum-ɔle of fallen* Neocalamites *trunks ınd dense fern and cycadophyte ınderstory to scavenge the rotting carcass of an oddly shaped aeto-aur,* Desmatosuchus. *The bony ɔlates of the armored and spiky ıetosaur made it a difficult ani-nal to kill and a tough animal o eat but the small dinosaurs vere able to use their razor-harp teeth to tear off the tail ınd to rip open the underbelly. Hungry but patient, the pack vill bide its time and return for nore when the phytosaur is ɔone and the slough is safe. The ıge of dinosaurs has begun.*

chapter six

6

The Time of the First Dinosaurs:

The Triassic and the Jurassic

National Park, Arizona

[Today]

Petrified Forest National Park is a series of broken patches of lavender, red, pastel green, and gray badlands and desert scrublands in northeastern Arizona. The multicolored badlands expose the Chinle Formation, rocks that were deposited as mud and sand just north of the equator about 225 million years ago. The park is named for the many giant petrified logs of long-extinct conifer trees that are strewn across the landscape.

The sedimentary rocks of the Chinle Formation hold many clues to their origin. Brightly colored mudstone layers represent soils that formed on ancient landscapes. Erosion of this fossil soil often exposes the upright trunks of *Neocalamites*, a giant relative of the horsetail plant that apparently grew in crowded thickets. Sandstone layers contain the ripple marks and telltale cross-beds of underwater sandbars of ancient streams.

Two main types of plant fossil preservation are found here: petrifactions of fossil wood and compressions of leaves and stems. Petrification occurs when mineral-laden waters percolate through the ground and deposit minerals in buried plant debris. In time, much of the original plant material is replaced by silica and other minerals; the resulting petrifaction is a near-perfect reproduction of the original plant. But some of the original organic cell structure remains. Scientists compare the cell structure of the fossil with that of modern plants to determine the nearest living relative. The most common wood type in the park is named

Araucarioxylon, and the microscopic detail of this fossil wood is similar to that of the living monkey-puzzle coniferous trees of the Southern Hemisphere. From the size of the petrified logs, we can determine that the trees reached tremendous heights, perhaps 200 feet. Other than *Neocalamites*, there are very few standing trunks in the Chinle Formation, and it appears that many of the trees did not grow on the site but floated in from somewhere else. This fossil wood really represents more of a petrified logjam than a petrified forest. The formation of compression fossils is easier to understand. These fossils are formed by the burial and flattening of leaves and twigs. Certain mudstone layers in the park preserve compression fossils of fern fronds, cycadophyte leaves, and conifer needles.

The Chinle Formation has also yielded a rich variety of vertebrate fossils including large amphibians, phytosaurs, aetosaurs, and dinosaurs. Phytosaurs were the large meat-eaters of inland waters. They resembled their relatives, the crocodiles, in appearance and lifestyle but had their nostrils in the mid portion of their skulls behind their snouts, allowing them to breathe with only their eyes and nostrils above water. Herbivorous aetosaurs like the 12-foot-long *Desmatosuchus* had large, sharp, and bony spikes projecting from its armadillo-like armor. Some of the earliest known dinosaurs are found in these rocks. Although small by dinosaur standards, the 3-foot-tall theropod *Coelophysis* was swift, agile, and probably very aggressive. It had small blade-like teeth that enabled it to inflict jagged wounds. In 1947, Edwin Colbert, a paleontologist at the American Museum of Natural History, discovered a *Coelophysis* bone bed near Ghost Ranch in north-central New Mexico. The close association of dozens of skeletons at this site suggests that *Coelophysis* may have lived and hunted in packs.

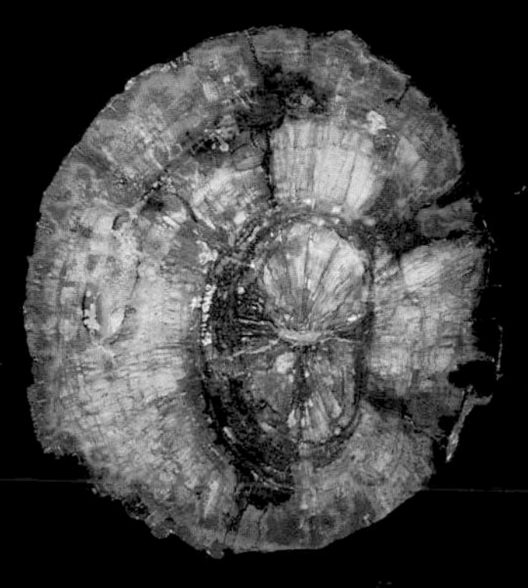

Petrified log

[Araucarioxylon arizonicum]

The brilliant colors of this petrified wood formed as mineral-laden groundwater percolated through the buried log.

Late Triassic, 225 mya
Chinle Formation, Arizona
Log diameter: 17.1 inches, DMNH 2666

Cycadophyte leaf

[Zamites powelli]

Cycads and cycad-like bennetitaleans were abundant plants in the late Triassic period.

Late Triassic, 225 mya
Chinle Formation, Arizona
Slab length: 11.2 inches, DMNH 7884

Theropod dinosaurs

[Coelophysis bauri]

Delicate and graceful, this pair of *Coelophysis* shows the bipedal stance characteristic of early dinosaurs. Here, a juvenile playfully chases an adult, trying to take a bite of the dangling aetosaur tail.

Late Triassic, 225 mya Chinle Formation, New Mexico
Length: 7.8 feet, DMNH 14729
Length: 4.9 feet, DMNH 22702

6 chapter six

The Time of the First Dinosaurs:

The Triassic and the Jurassic

The Triassic World

[225 MILLION YEARS AGO]

The interior of the supercontinent Pangaea is alternately subjected to giant monsoonal rains and to dry periods.

The age of reptiles: The Mesozoic

The age of reptiles—the Mesozoic era—commenced about 250 million years ago. The first periods of the Mesozoic, the Triassic and the Jurassic, witnessed major changes in both the land flora and fauna. The gymnosperms became the dominant plants of the Mesozoic. Protomammals and primitive reptiles were the common animals of the land faunas at the beginning of the Mesozoic. But the gigantic sauropod dinosaurs and their kin, the carnosaurs, and the ornithischian dinosaurs, soon replaced them. Among the reptiles, many major groups including the dinosaurs, pterosaurs, lizards, turtles, and marine ichthyosaurs and plesiosaurs appeared in the Triassic, as did the modern frogs and salamanders. Coincident with the origin of the dinosaurs is the origin of our own class of vertebrates—mammals—about 225 million years ago. The first birds appear in the Jurassic and are represented by *Archaeopteryx* from Europe and other fossils from China.

The early Mesozoic earth was vastly different from today. The supercontinent Pangaea formed during the Permian around 250 million years ago by the collision of all the world's continents, and it persisted into the Jurassic. This large supercontinent extended symmetrically across the equator from pole to pole. At about this time, Petrified Forest National Park was near the equator, about 400 miles east of the ocean that bordered the supercontinent. In stark contrast to the desert environment in this region today, the climate was monsoonal with extremes in rainfall. Coal deposition during the Triassic occurred at high southern and northern latitudes, which suggests that these areas were wetter than the interior of Pangaea. The breakup of Pangaea began about 170 million years ago, and our current continents and the Atlantic Ocean began to take shape.

The dinosaur dynasty

Everyone is fascinated by dinosaurs—the animals that dominated ecosystems on land for about 160 million years. From TV programs to advertisements to Hollywood movies and educational programming, dinosaurs are a part of our lives. Both children and adults are attracted to these creatures. The dinosaur names *Triceratops, Tyrannosaurus,* and *Stegosaurus* are more familiar to young children than are the names of the presidents of the United States.

Dinosaurs have been depicted as both vicious and gentle and as both large and small; but there never really was a typical dinosaur. They ranged in size from that of a chicken to a locomotive. All hatchling dinosaurs were smaller than a newborn human baby; some were as small as a mouse. Elaborate head gear and body armor adorned many dinosaurs, and it is quite likely that they were as colorful as today's tropical birds. Most of the dinosaurs were very specialized ecologically, often feeding exclusively on plant parts or on flesh.

We now have a completely new view of the metabolism and behavior of dinosaurs. Evidence has come not only from the skeletal remains of these magnificent creatures but also from their nesting sites and the trackways they left behind on ancient muddy surfaces. Studies of bone growth have shown that dinosaur metabolism may have been warm-blooded and different from that of other reptiles, and the fact that dinosaurs are closely related to birds supports this idea. The skeletons of all dinosaurs show that their limbs were directly under their bodies. This position is more efficient than that of reptiles such as lizards, in which the limbs project out from the sides of the body. The configuration of dinosaur limbs allowed for swift movement, and an animal such as *Coelophysis* presents a skeletal impression of great agility and movement.

Some dinosaurs probably had complex social patterns that rivaled those of most modern mammals and birds.

Gingko leaf

[Sphenobaiera sp.]

Fan-shaped leaves are character istic of ginkgo trees. Ginkgos and their relatives have existed since the Triassic period.

Late Triassic, 215 mya
Narabeen Group, Australia
Slab length: 8.3 inches, DMNH 7880

Many developed sophisticated social groups. Some herbivorous dinosaurs lived in herds and migrated over vast distances together. Carnivorous dinosaurs also may have hunted in packs. Perhaps the most fascinating aspect of dinosaur behavior, according to new evidence from nesting sites discovered by Jack Horner of the Museum of the Rockies, is that parents may have cared for their young. Several generations may have lived together in large social groups near nesting areas.

Trackways of dinosaurs have been discovered in almost every part of the globe. Some of these trackways support the idea that even such large creatures as the sauropods may have traveled in groups and others suggest that the theropods were able to run quite swiftly. It is quite clear that dinosaurs were not slow, dumb creatures but agile and active ones.

The oldest dinosaur fossils (*Herrerasaurus* and *Eoraptor*) are found in Triassic rocks in Argentina that are 228 million years old. These dinosaurs were relatively small and walked on their two hind legs. Their skulls were typical of later dinosaurs, and they had sharp teeth, indicating that they were carnivorous.

The dinosaurs split into two major groups during the Triassic: the ornithischians and the saurischians. The ornithischians were herbivorous dinosaurs with a hip similar to that of modern birds. Their pubis bone extended below and parallel to the ischium as in modern birds. The ornithischians included a number of herbivorous dinosaurs, such as the well-known hadrosaurs, ceratopsians, and stegosaurs; the armored nodosaurs; and the dome-headed pachycephalosaurs. Many of these groups developed elaborate head ornamentation and horns as well as body armor. The ornithischians were not as common as the saurischians in the Triassic and the Jurassic. They eventually dominated dinosaur communities by the end of the Mesozoic, during the Cretaceous period.

The saurischians, the so-called lizard-hipped dinosaurs, included carnivorous, omnivorous, and herbivorous dinosaurs. They had a pelvic structure in which the pubis bone usually extended directly toward the front. One group of saurischians—the theropods—included small to large carnivores such as *Ceratosaurus* and *Allosaurus*. Another

Conifer cone

[Araucaria mirabilis]

This petrified cone preserves both its outer shape and texture and its internal structure.

Middle Jurassic, 160 mya
Cerro Cuadrado Formation, Argentina
Cone length: 2.8 inches, DMNH 5904

Fern frond

[Cladophlebis heterophylla]

Ferns were common herbaceous plants in the Jurassic and may have been fodder for the giant dinosaurs.

Late Jurassic, 150 mya
Morrison Formation, Montana
Slab length: 4.7 inches, DMNH 7820

Sauropod dinosaur

Denver Museum's Diplodocus *is 80 feet long. The 50-foot-long whip-like tail is long even by sauropod standards and was probably used to ward off predators.*

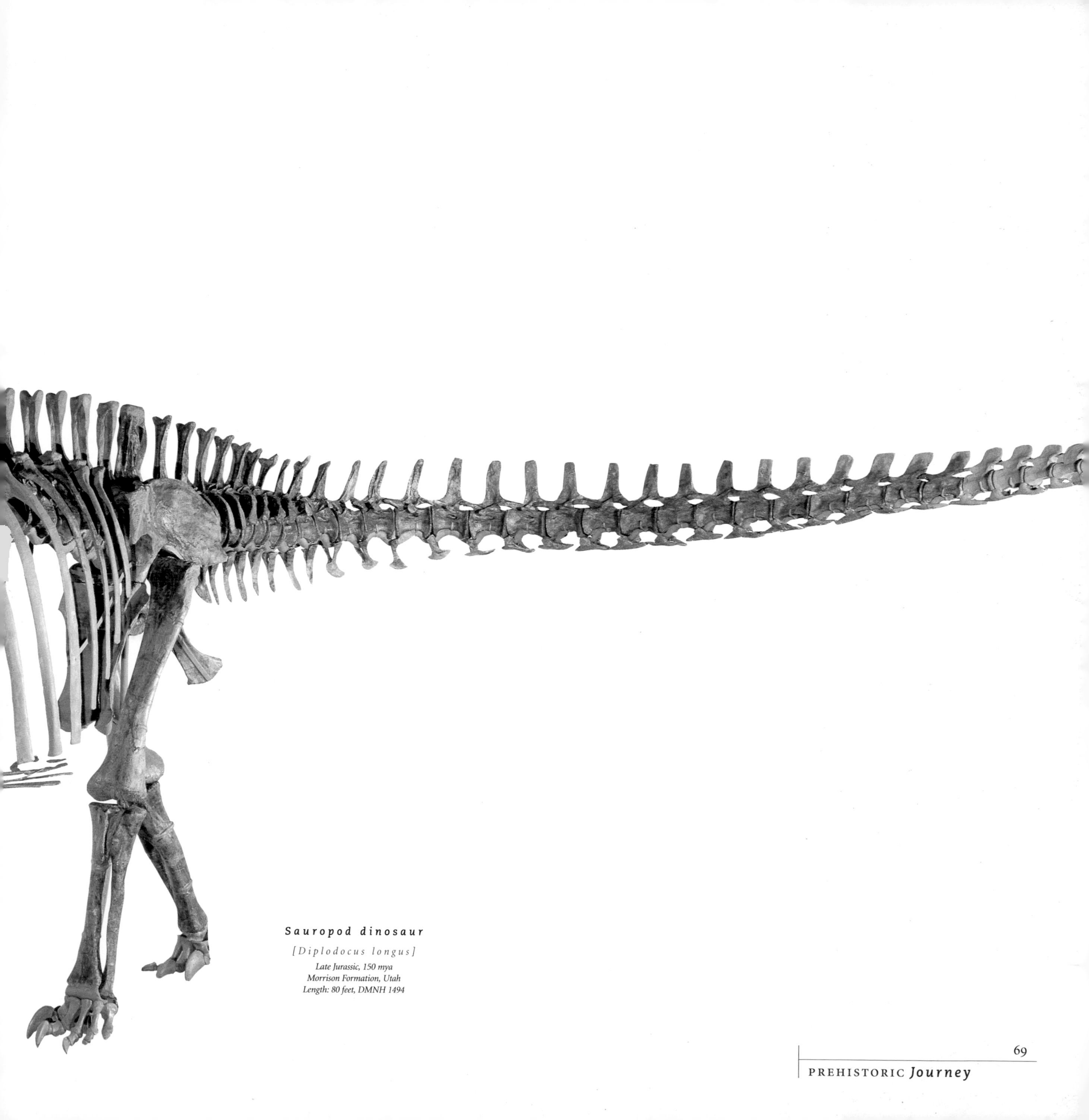

Sauropod dinosaur

[Diplodocus longus]

Late Jurassic, 150 mya
Morrison Formation, Utah
Length: 80 feet, DMNH 1494

A Jurassic meadow

[DIPLODOCUS]

Eighty feet and 20 tons of sinuous and stately grace, a Diplodocus *prances through a waist-high meadow of ferns and cycads on a warm spring day in Colorado, 150 million years ago.*

group—the sauropodomorphs—included both omnivores and herbivores. Sauropodomorphs were often massive in size and were the largest land animals of all time. In their heyday during the late Jurassic, these giants included *Brachiosaurus, Diplodocus, Ultrasauros, Apatosaurus* (Brontosaurus), and *Supersaurus.*

Dinosaur egg clutch

[Prismatoolithus coloradensis]

Dinosaur nests were first discovered in the 1920s in Mongolia. Since then they have been found at many other places in the world. This nest from Colorado is the oldest one known from North America.

Late Jurassic, 150 mya
Morrison Formation, Colorado
Nest length: 1.6 feet, DMNH 21679

Dinosaurs of the Morrison Formation

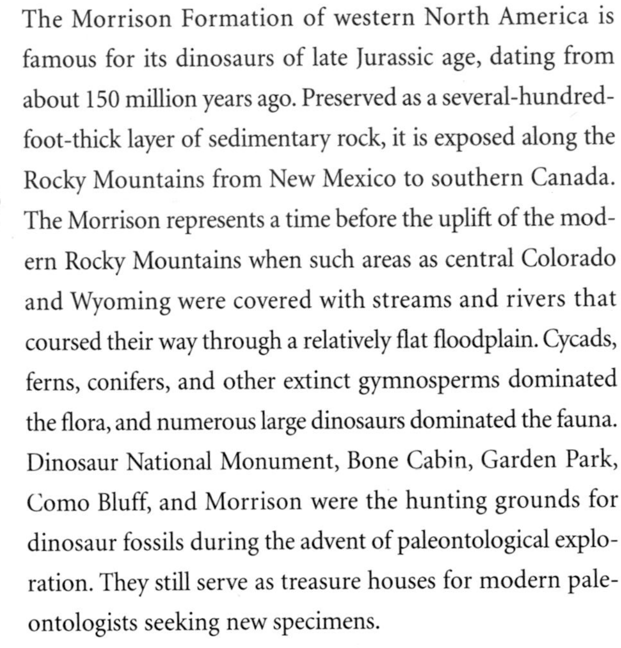

The Morrison Formation of western North America is famous for its dinosaurs of late Jurassic age, dating from about 150 million years ago. Preserved as a several-hundred-foot-thick layer of sedimentary rock, it is exposed along the Rocky Mountains from New Mexico to southern Canada. The Morrison represents a time before the uplift of the modern Rocky Mountains when such areas as central Colorado and Wyoming were covered with streams and rivers that coursed their way through a relatively flat floodplain. Cycads, ferns, conifers, and other extinct gymnosperms dominated the flora, and numerous large dinosaurs dominated the fauna. Dinosaur National Monument, Bone Cabin, Garden Park, Como Bluff, and Morrison were the hunting grounds for dinosaur fossils during the advent of paleontological exploration. They still serve as treasure houses for modern paleontologists seeking new specimens.

The first dinosaur specimen discovered in western North America was described in 1870 from a single tail vertebra from the Morrison of Middle Park, Colorado. It was named *Antrodemus* by Joseph Leidy, a paleontologist from the Philadelphia Academy of Natural Sciences. Only seven years later, in 1877, the great dinosaur "gold rush" of the West began with two of the most famous paleontologists of the nineteenth century. Othniel Charles Marsh of Yale College and Edward Drinker Cope of the University of Pennsylvania were told that large bones had been discovered along the Front Range of the Rocky Mountains in deposits near Morrison west of Denver and at Garden Park near Cañon City in southern Colorado. More discoveries came at the site of Como Bluff in Wyoming in 1879. These finds included the great sauropods, stegosaurs, and other dinosaurs. What began as a congenial professional relationship between Marsh and Cope in the quest to understand the history of life turned into a rivalry. The rivalry turned bitter as both of them tried to outdo the other in naming new animals. Publication of findings was often rapid but premature, and inevitable redundancies in naming similar animals resulted. One famous dinosaur was named twice. In 1877, Marsh described *Apatosaurus* from bones discovered near Morrison. In 1879, he named another dinosaur Brontosaurus from bones discovered at Como Bluff, Wyoming. Eventually it was determined that they were the same animal! Because *Apatosaurus* was named first, the scientific convention is to use this name instead of the better known Brontosaurus.

The early discoveries by Marsh and Cope in Colorado and later in Wyoming provided a vivid picture of Mesozoic life. By the end of the nineteenth century, *Stegosaurus, Camarasaurus, Diplodocus, Ceratosaurus,* and *Allosaurus* were fleshed out into artistic reconstructions that still typify the age of reptiles. In addition to dinosaurs, the Morrison Formation has yielded a diverse fauna of crocodiles, turtles, and lizards, and many primitive mouse-size mammals. Agatized freshwater snails also abound in some layers of the Morrison.

One of the most spectacular sites of Morrison dinosaurs, now known as Dinosaur National Monument, is on the northern Colorado-Utah border; it was found by Earl Douglass of the Carnegie Museum of Natural History in 1909. Through 1922 the Carnegie Museum excavated this site. From it came more than 5,000 bones representing 60 individuals of 10 different dinosaur species including *Diplodocus, Camarasaurus, Apatosaurus, Allosaurus, Dryosaurus, Camptosaurus,* and *Stegosaurus.* The largest land animals to have ever walked the earth were the giant sauropods from the Morrison, including *Brachiosaurus, Ultrasauros,* and *Supersaurus* from sites in western Colorado and *Seismosaurus* from later deposits in New Mexico. *Brachiosaurus* stood over 50 feet tall, was nearly 80 feet long, and weighed more than 80 tons, while *Seismosaurus,* with an estimated length of 120 feet, may have been the largest land animal ever.

Freshwater snails

Red, pink, and polka-dotted, these tiny snails lived in shallow streams and lakes frequented by huge dinosaurs. The colors formed as the snail shells were replaced by silica during fossilization.

Freshwater snails

[Viviparus reesidei]

Late Jurassic, 150 mya
Morrison Formation, Colorado
Length: 11.4 inches, DMNH 6003

Struggle for survival

One of the best known predators of the Jurassic, Allosaurus *was smaller than* Tyrannosaurus rex *and differed from it in having three fingers on its short, arm-like front leg.* Stegosaurus *had broad plates running along its back, an armored throat, and nasty tail spikes to protect it from such predators as* Allosaurus.

Allosaurus and Stegosaurus

[Allosaurus fragilis and *Stegosaurus stenops]*

Late Jurassic, 150 mya
Morrison Formation, Colorado
Length: 20 feet, DMNH 2149
Length: 19 feet, DMNH 1483

Stegosaur head and tail

[Stegosaurus stenops]

A complete *Stegosaurus* skeleton, discovered in 1992 by Bryan Small of the Denver Museum of Natural History, demonstrated for the first time that the neck region was protected by a collar of bony ossicles and that the tail spikes were horizontally attached.

Late Jurassic, 150 mya
Morrison Formation, Colorado
Skull length: 17.7 inches
Tail length: 7 feet
DMNH 2818

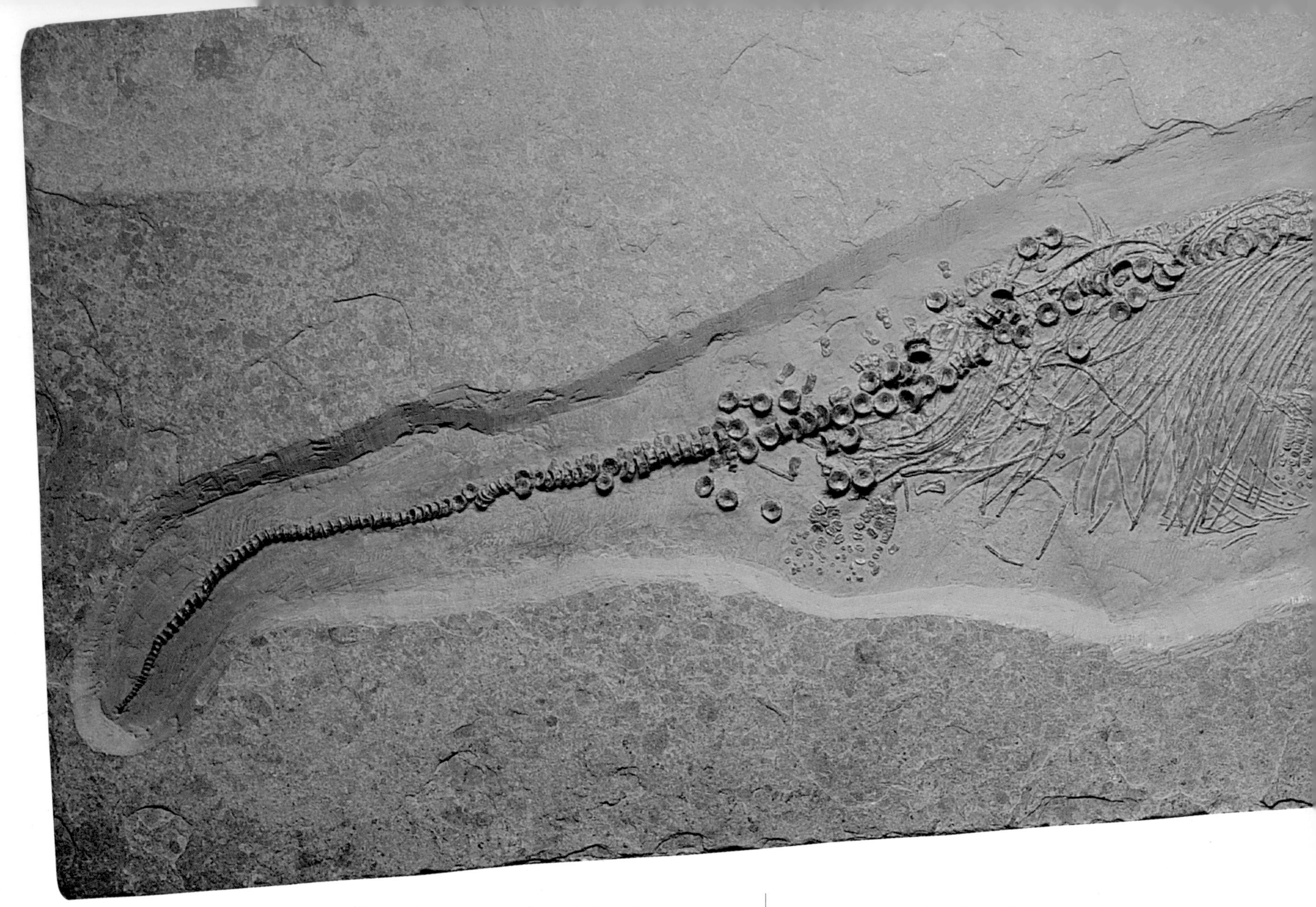

Discoveries continue to be made from the Morrison. A specimen of *Allosaurus fragilis* was discovered in 1979 in Moffat County, Colorado, by India Wood, a 13-year-old schoolgirl. She donated it to the Denver Museum of Natural History in 1982. The specimen preserves about 50 percent of the bones. In 1992, the most complete skeleton of *Stegosaurus* was discovered in Garden Park by a Denver Museum expedition led by Kenneth Carpenter and Bryan Small. This new specimen solves many of the mysteries surrounding the plate arrangement and tail of *Stegosaurus* and adds a new quirk to this enigmatic dinosaur's armor. The new specimen shows that the plates on the back overlapped one another and the tail was held up in the air with the spikes pointing out to the sides. A surprise discovery was that hundreds of small hexagonal bony pieces were closely packed in the neck region of the *Stegosaurus,* providing a natural chain-mail armor across its chest.

Marine reptiles, pterosaurs, birds, and early mammals

Although dinosaurs are the hallmark of the Mesozoic, many other vertebrates made their first appearance in the early part of the era. Salamanders, frogs, crocodilians, and turtles appeared in the early Mesozoic; even at this stage, they were quite similar to their modern counterparts. Early

Ichthyosaur

[Stenopterygius quadriscussis]

Ichthyosaurs were dolphin-like reptiles that were entirely aquatic. The ichthyosaurs were the dolphins of the Mesozoic seas.

Early Jurassic, 185 mya
Posidonia Shale, Holzmaden, Germany
Skeleton length: 4.9 feet, DMNH 2729

Nothosaur

[Pachypleurosaurus staubi]

This small reptile, a relative of the plesiosaurs, lived primarily in water but also had the ability to move about on land.

Middle Triassic, 230 mya
Santa Rosa Formation, Italy
Skeleton length: 8.9 inches, DMNH 2735

sphenodontids, relatives of the modern tuatara of New Zealand, and animals related to today's glass lizards and skinks also appear in Jurassic deposits.

Beyond the giant proportions reached by dinosaurs, perhaps the most remarkable adaptations of reptiles during the Mesozoic were those that allowed them to live in the sea. The late Triassic *Pachypleurosaurus* was a miniature precursor to the plesiosaurs, which dominated the oceans during the Jurassic and Cretaceous. *Pachypleurosaurus* was small and short-limbed and still was able to live on land. Its feet with multiple bones in each digit do, however, suggest of a life in the water. In contrast, plesiosaurs were fully aquatic. They often possessed long necks, short tails, and long flippers. Plesiosaurs were graceful swimmers who traveled through the water much like some modern aquatic birds (penguins, for example).

The ichthyosaurs, or fish-like reptiles, are quite common in some deposits of Jurassic age. They had fish-like bodies, long snouts, and flippers that made for very efficient swimming. A remarkable fossil of a Jurassic ichthyosaur from Holzmaden, Germany, preserves the mother ichthyosaur in the act of giving live birth, a feature of an animal fully adapted to marine life. Food resources were plentiful in the Jurassic seas. Coiled ammonites and bullet-shaped

Ammonites

The ammonites on this slab were partially filled with mud when they were buried at the bottom of a Jurassic sea. Much later, percolating groundwater deposited yellow calcite crystals in the hollow shells.

Ammonites

[*Promicroceras planicosta* and *Asteroceras obtusum*]

Early Jurassic, 195 mya
Lower Lias Formation, England
Slab length: 1.8 feet, DMNH 6170

belemnites are two types of cephalopods that are extremely common fossils in Jurassic marine rocks, sometimes accumulating in mass death horizons.

Two groups of vertebrates took to the air in the Mesozoic: pterosaurs and birds. Both groups appeared late in the Triassic period. Pterosaurs were an extremely diverse group of flying reptiles. Many were fish-eaters, but some evidence suggests that others may have eaten small oceanic planktonic microorganisms and yet others fed on hard-shelled invertebrates. Some pterosaurs had hair and most ranged in size on the scale of modern seabirds. *Quetzalcoatlus* of the Cretaceous was immense. With a wingspan of 50 feet, it is the largest animal to have ever flown over the earth.

Archaeopteryx, presumably the earliest known bird, is identified by a feature unique to birds: feathers. *Archaeopteryx* had three separate clawed fingers and possessed a wishbone like modern birds. But unlike modern birds, it had teeth. The origin of birds has been debated since the discovery of *Archaeopteryx*. Charles Darwin's *Origin of Species* appeared in 1859, only two years before *Archaeopteryx* was discovered. *Archaeopteryx* was the first in a long line of "missing links" that fill in the gaps between ancient creatures and their modern descendants. Currently, paleontologists favor the idea that birds are descended from a saurischian dinosaur, perhaps similar to *Compsognathus* or one of the advanced theropod groups such as the carnivorous dromaeosaurs. *Archaeopteryx,* in fact, is so similar to *Compsognathus* that for many years a specimen of the bird had been labeled with the name of the dinosaur.

The origin of flight among birds is also a topic of great debate. Some claim that bird flight evolved when a ground-dwelling ancestor ran and became airborne. Others suggest that early bird ancestors lived in trees and began flight by gliding. Whichever hypothesis one favors, *Archaeopteryx* was fully capable of flight and possessed fine delicate bones and feathers. By the end of the Mesozoic, birds had evolved into a number of different groups, but much of their diversification occurred only after the dinosaurs became extinct.

Mammals first appear in the Triassic at the same time as dinosaurs, 225 million years ago. For the first two-thirds of their entire history, mammals were a group of small, secretive, insect- and seed-eating animals that inhabited the forest undergrowth. Mammals evolved from synapsid protomammals through a series of stages that resulted in the basic characteristics of all living mammals. The jaw musculature was enhanced through the development of bony struts on the skull. The bones that support the jaw on the skull were reduced in size and incorporated into the middle ear, which typifies the mammals. The mammals also developed four kinds of teeth—incisors, canines, premolars, and molars. Although most reptiles replace their teeth throughout their lifetime, mammals only had two sets of teeth—the deciduous, or baby, teeth and the adult teeth.

The earliest mammal, *Adelobasileus*, is known from only the back of the skull and was found in late Triassic rocks in Texas. One of the earliest mammals, *Morganucodon* from the late Triassic and early Jurassic, was only 4 inches long and weighed about 1 ounce. The skeleton of *Morganucodon* suggests that it was able to climb, scramble, and grasp with equal agility. Both of these early mammals were more advanced than any protomammal with specializations in the skull and ear. At the end of the Jurassic, there was still an 80-million-year countdown until the end of the age of reptiles.

Mammal jawbone

[Docodon sp.]

Mammals appeared at the same time as the dinosaurs 225 million years ago but remained small until the dinosaurs became extinct 65 million years ago.

Late Jurassic, 150 mya
Morrison Formation, Colorado
Jaw length: 1.1 inches, DMNH 14730

Marmarth, North Dakota

[66 Million Years Ago]

In a dank streambed, two bear-sized, bipedal Stygimoloch *dinosaurs lunge at one another, each trying to gouge its adversary's broad flank with its thick and spiky head. Other than this bizarre pugilistic encounter, the scene seems familiar. Vine-draped, broad-leaf trees cast dappled shadows across the muddy streambed. Palm trees lend a subtropical look. Bird songs and the buzz of flying insects fill the air. A putrid smell arises from something partially buried by the shallow stream—a huge, rotting skull with three sharp protruding horns that belonged to a recently vital* Triceratops. *Beyond the skull lies the scattered remnant of the* Triceratops' *rhino-sized carcass. Gigantic footprints, testimony to the recent passage of a* Tyrannosaurus rex, *have filled with bits of debris. Above the streambank, cycads, ground ferns, and leaf litter cover the forest floor. An aggressive marsupial peers from behind a tree. Overhead, a pterosaur glides through the forest. In the distance, an ostrich-like* Ornithomimus *browses on cycad leaves while a herd of hadrosaurs works its way through the forest on the far side of the river.*

chapter SEVEN

The Last of The Dinosaurs:

The Cretaceous

Marmarth, the southwesternmost town in North Dakota, was named by a man who couldn't decide which of his granddaughter's names to use: Martha or Margaret. Once a bustling cattle-shipping town on the Milwaukee Railroad, Marmarth today is a sleepy shadow of its former self. Set deep in the wide valley of the Little Missouri River, the town is surrounded by the somber gray and brown hills of the Hell Creek Formation. The Little Missouri is a small river, as wide as a city street and rarely deeper than a mud puddle for most of the summer. But over the past 10,000 years it has been active, carving a valley 5 miles wide and 500 feet deep and exposing hundreds of square miles of upper Cretaceous rock. Composed of sediments that have just barely turned to stone, the Hell Creek weathers into banded badlands that look like piles of melting ice cream. And these hills do melt. Each rainfall washes away more sediment and exposes new fossils.

The Hell Creek Formation is famous for its dinosaurs. It was the Hell Creek in central Montana that in 1903 yielded the world's first *Tyrannosaurus rex. Triceratops* skulls are so common in these rocks that Robert Bakker, Colorado's popular dinosaur scientist, calls these animals the cockroaches of the Cretaceous. At one spot in central South Dakota, a bone bed composed almost entirely of duck-billed hadrosaurs stretches for hundreds of yards and contains the remains of hundreds of animals. The Hell Creek Formation contains nearly 20 species of dinosaurs including the poorly understood *Stygimoloch*

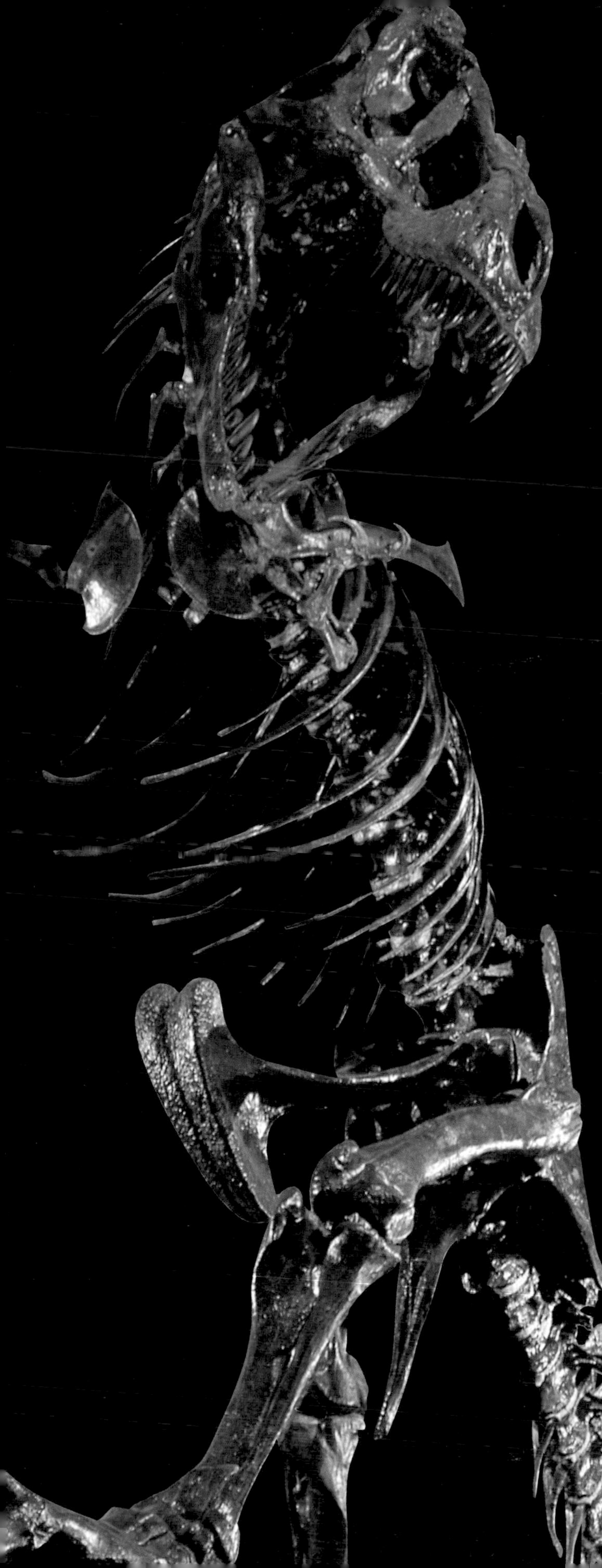

T. rex

[*Tyrannosaurus rex*]

Tyrannosaurus rex, first collected from the Hell Creek Formation is the largest terrestrial carnivore of all time.

Late Cretaceous, 66 mya
Hell Creek Formation, Montana
Skeleton length: 40 feet, DMNH 2151

spinifer (the name means spiny devil from the river Styx). Known only from two partial skulls collected in the 1980s, *Stygimoloch* is one of the bone-headed, or pachycephalosaur, dinosaurs.

The Cretaceous-Tertiary boundary that lies at the top of the Hell Creek Formation represents the time of the extinction of the dinosaurs. Near Marmarth, the formation is a 350-foot-thick section of sedimentary rocks formed from the deposits of meandering rivers and associated floodplains laid down between 67 and 65 million years ago. The geometry of the sand and mud layers is a clue to the width, depth, and flow direction of the ancient stream channels. More than just a dinosaur graveyard, these rocks contain accumulations of tiny bone and jaw fragments of small animals such as lizards, turtles, snakes, birds, amphibians, mammals, small dinosaurs, crocodiles, and alligator-like champsosaurs. Fossil roots and trunks identify the location and nature of ancient forests. Fossil leaves, flowers, and pollen betray the nature of the vegetation and the climate in which it grew. Over 90 percent of the more than 230 species of Hell Creek fossil leaves came from broad-leafed trees and shrubs. Many of the leaves bear distinctive marks of insect damage and, occasionally, even insect fossils are found. This formation provides an unusually clear view of the waning moments of the Mesozoic era.

Sycamore leaf

[*Erlingdorfia montana*]

By the end of the Cretaceous period, flowering plants were common in most ecosystems.

Late Cretaceous, 66 mya
Hell Creek Formation, North Dakota
Slab length: 6.3 inches, DMNH 6253

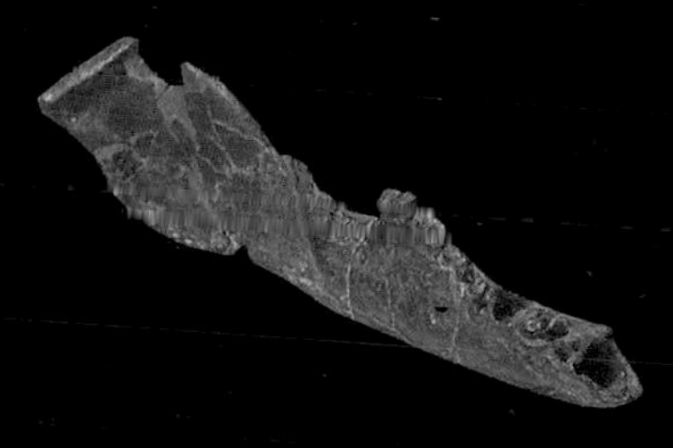

Marsupial jawbone

[*Didelphodon vorax*]

In the late Cretaceous period marsupials were among the most common mammals in North America.

Late Cretaceous, 66 mya
Lance Formation, Wyoming
Length: 3.5 inches, DMNH 4327

7 chapter SEVEN

The Last of the Dinosaurs: The Cretaceous

The Cretaceous World

[120 MILLION YEARS AGO]

The supercontinent Pangaea splits into two parts: Laurasia and Gondwana. Global climate is warm and many continents are flooded by shallow seas.

Conifer needles

[*Araucarites* sp.]

These conifer needles were buried by volcanic ash about 81 million years ago.

Late Cretaceous, 81 mya
Rock Springs Formation, Wyoming
Branch length: 4.3 inches, DMNH 5917

Life on land and the rise of the flowering plants

The Cretaceous period began about 144 million years ago and lasted until about 65 million years ago. When it began, earth was dominated by ferns, gymnosperms, dinosaurs, and marine and flying reptiles. The characteristic creatures of the modern world were to make their appearance during the Cretaceous, and the extinction of the giant reptiles would open the way for the formation of the modern ecosystems. It has been said that if you took a random grab sample of the world today, you would likely get a beetle and an herbaceous flowering plant, or angiosperm. If you were to find a vertebrate animal, it would likely be a mammal, bird, or fish. Much of the early history of these groups is embedded in the Cretaceous.

Perhaps the most dramatic change to occur in the Cretaceous was the proliferation of the angiosperms. Angiosperms, which today represent about 80 percent of all 250,000 known species of land plants, appeared and quickly diversified in a world previously dominated by ferns and gymnosperms. As a group, the angiosperms are incredibly versatile; their members include giant rainforest trees, grasses, tundra herbs, palms, wildflowers, sea grasses, and fruit trees. Flowering plants are the major element in most terrestrial ecosystems, making up the bulk of the biomass of the green planet. The earliest known angiosperms appear in the fossil record as pollen grains from equatorial sites about 125 million years ago. Within 50 million years, the angiosperms had evolved to dominate the vegetation of the earth. The explosive success of this plant group has much to do with its partnership with the animal kingdom. To achieve pollination and reproductive success, flowering plants interact with insects, birds, and mammals; in addition, they take advantage of the tried-and-true gymnosperm techniques of wind and luck. Once the fertilized flower ripens into a seed-bearing fruit, it is dispersed by birds, mammals, reptiles, and wind.

Insects are the animal equivalent to the flowering plants. With more than 1 million named species and perhaps as many as 10 to 30 million species awaiting discovery in our rapidly disappearing rainforests, insects are the most abundant life form on the planet. Because of their small size, rapid generation time, and mobility, insects evolved quickly. Recent compilations of the insect fossil record show that insects have been steadily increasing in diversity since the Permian extinction that knocked out a few of the early groups. Although it has long been thought that the insects and the angiosperms helped each other by coevolution, there does not appear to be a burst of insect evolution coincident with the late Cretaceous radiation of the angiosperms.

Although the early interaction of insects and flowers is unclear, the relationship between plants and larger herbivores is clearer. The dinosaurian fauna of the Jurassic and early Cretaceous were dominated by giant, long-necked sauropods, animals capable of foraging high above the ground for vegetable matter that they then processed in gastric mills. These animals had simple teeth for cropping plants but apparently did not chew their food. The herbivorous dinosaurs of the late Cretaceous were fundamentally different. They had large heads and powerful jaws. The hadrosaurs had duck-like bills in front of batteries of teeth that could chew their food to a pulp. Ceratopsians had sharp bills and rows of shearing teeth. Hadrosaurs were much smaller than sauropods, and their effective foraging range was much more limited. The ceratopsians, who carried their heads very near the ground, like modern grazing animals such as bison, had an even more limited feeding range. The evolution of new kinds of herbivorous dinosaurs with different types of mouths and teeth speaks to the appearance of new food sources. These were almost surely the angiosperms.

Plants of the early Mesozoic

[A WORLD WITHOUT FLOWERS]

The world 130 million years ago was truly a green planet. Conifers, ferns, cycads, cycadeoids, ginkgos, horsetails, and mosses covered a world without flowers.

With the new herbivorous dinosaurs rose groups of new carnivorous animals, including the giant tyrannosaurids and the lesser dromaeosaurs, troodontids, and velociraptors. *Deinonychus,* a small predatory dinosaur discovered in Wyoming in 1966, inspired Yale University paleontologist John Ostrom to kick off the dinosaur renaissance and to suggest that some dinosaurs were warm-blooded.

Although late Cretaceous dinosaurs have been found in many places around the world, the most complete record comes from the Western Interior of North America. Recent finds in central Asia, southern South America, and Europe help to flesh out the record. In North America, sauropods disappear during the early Cretaceous but reappear 70 million years later near the end of the late Cretaceous. Late Cretaceous deposits in New Mexico contain a sauropod, known as *Alamosaurus,* that apparently migrated into North America from South America.

When Cretaceous seas covered the continent

While dinosaurs roamed the land, a variety of marine reptiles shared the sea with a diverse assemblage of sharks, bony fish, and ammonites. Large predators were common and the seas were not a safe place. The best known fossils of the Cretaceous seas are the ammonites, shelled cousins of the squid, cuttlefish, and chambered nautilus. Present since the Devonian and survivors of many bouts of extinction, the ammonites reached their maximum diversity in the Cretaceous but failed to survive it.

For much of the late Cretaceous, the Western Interior of North America was covered by a shallow sea. Sometimes as wide as 2,000 miles and several hundred feet deep, this seaway connected the Gulf of Mexico with the Arctic Ocean and effectively split North America into two continents. You would have needed an ocean liner to get from Salt Lake City to St. Louis. This seaway was bounded by mountain ranges on the west and low-lying land on the east. The sea effectively covered the middle of North America for the last 40 million years of the Cretaceous and deposited marine sediment as much as 4 miles thick.

The source of the sediment was the mountains along the west coast of the sea in what is now Utah, Nevada, and Idaho. Near the mountains, coarse gravels were left by rapidly flowing streams. Near the coast, sand and mud were deposited by meandering rivers and thick layers of peat formed in coal swamps. The coasts were characterized by sandy beach deposits and the layered muds of tidal

Rose Creek flower

[Saxifragales]

The five-petaled Rose Creek flower is one of the oldest recognizable fossil flowers. Fossil flowering plants are also recognized by their distinctive pollen, wood, leaves, and fruits.

Late Cretaceous, 95 mya
Dakota Formation, Nebraska
Flower width: 0.8 inch, DMNH 6028

Poppy fruit

[Palaeoaster inquirenda]

Fossil fruits are usually only preserved when the fruit has tough outer parts. This fruit, related to modern poppies, is a common fossil because its parts were woody and tough.

Late Cretaceous, 75 mya
Neslen Formation, Utah
Fruit length: 3.5 inches, DMNH 6333

Angiosperm leaf

[Leepierceia preartocarpoides]

Broad leaves with net veins are characteristic of many flowering plants. These types of fossils first began to appear about 130 million years ago.

Late Cretaceous, 66 mya
Hell Creek Formation, North Dakota
Slab length: 7.2 inches, DMNH 6359

A flowering of plants

[RISE OF THE ANGIOSPERMS]

In the early part of the Cretaceous period, flowering plants first evolved. Since then the world has been a colorful riot of plant diversity. Today, flowering plants are dominant in most ecosystems and are represented by more than 215,000 different species.

Duck-billed dinosaur

The duck-billed dinosaurs have traditionally been articulated in museum exhibits standing on their hind legs and tail like kangaroos. Evidence from trackways and the presence of ossified tendons along their spine show that they actually walked on all fours and held their rigid tails above the ground.

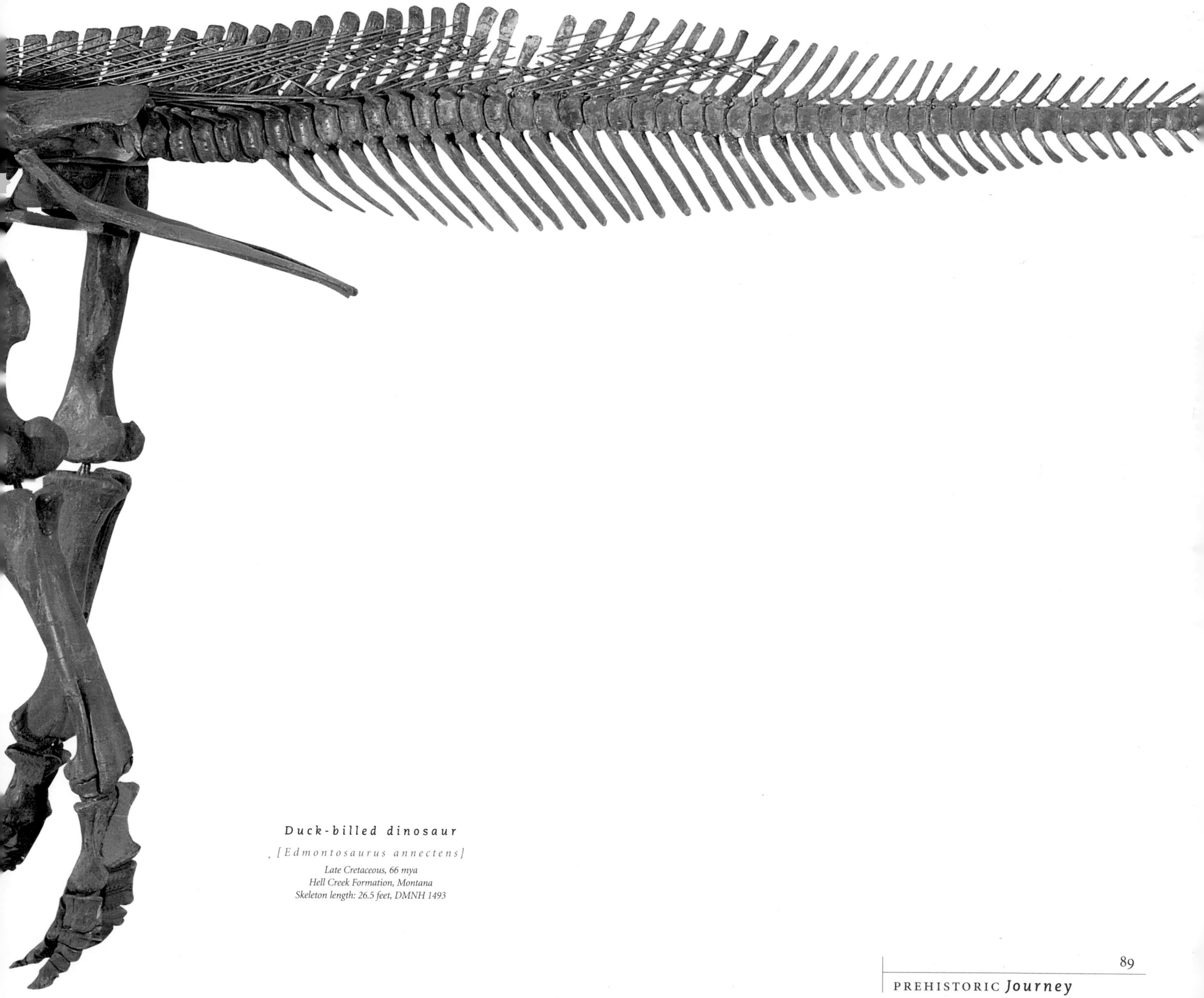

Duck-billed dinosaur

[Edmontosaurus annectens]

Late Cretaceous, 66 mya
Hell Creek Formation, Montana
Skeleton length: 26.5 feet, DMNH 1493

Duck-billed dinosaur

[EDMONTOSAURUS]

A duck-billed hadrosaur grazes in a late Cretaceous woodland in eastern Montana. Hadrosaurs were some of the last dinosaurs to walk the planet.

estuaries. Thin sand layers and stinky dark mud formed the shallow seafloor. To the east, beyond the influence of shoreline sediments, tiny marine invertebrates contributed their bodies to what would become the chalk beds of western Kansas.

The rocks that formed from the sediments that filled the ancient seaway are common sights in Colorado and the Rocky Mountain states today. These black shales and white limestones preserve many fossils of the animals that inhabited the seaway. The giant cliffs to the north of Grand Junction, Colorado, are composed of late Cretaceous marine mudstone of the Mancos Shale. The dry buttes here were once the bottom of a sea. Close inspection of these rocks reveals fossil fish teeth and bones, flattened ammonites, and beautiful slabs of floating crinoids. *Uintacrinus,* a crinoid without a stalk, was a beautiful animal that floated in mats, feeding through its long tentacle-like arms.

Rocks of the same age that are much farther east are composed not of nearshore muds but of offshore chalk and limestone. In western Kansas and eastern Colorado, chalk and limestone from the Niobrara and Greenhorn Formations contain an amazing diversity of ancient sea creatures: mosasaurs, plesiosaurs, sharks, giant fish, giant clams, pterosaurs, and diving birds. Several spectacular specimens of the giant fish *Xiphactinus* have been collected in western Kansas. Some of these fossils preserve the skeletons of smaller fish in their rib cages, direct evidence of their failure to digest their last meals. On the floor of this part of the seaway were colonies of huge inoceramid clams, some of which were almost 6 feet in diameter. Surprisingly, some of these giant clams are preserved with a variety of fish fossils between their shells. These fish lived within the clam when it was alive, using the giant shell as a feeding place that was sheltered from predators.

The mud that settled to the bottom of the Cretaceous seaway that covered the American West preserves one of the greatest examples of ammonite evolution in the world.

Cretaceous crab

[Avitelmessus grapsoideus]

This crab was preserved in a concretion that formed soon after the animal died and was buried. Since the concretion formed early and is very hard, it protected the fossil from being distorted.

Late Cretaceous, 70 mya
Ripley Formation, Mississippi
Specimen length: 10.2 inches, DMNH 7804

Floating crinoids

Unlike the stalked crinoids of the Paleozoic, these Cretaceous crinoids floated in mats on the sea surface. Their long feeding arms dangled down into the seawater, filtering out plankton.

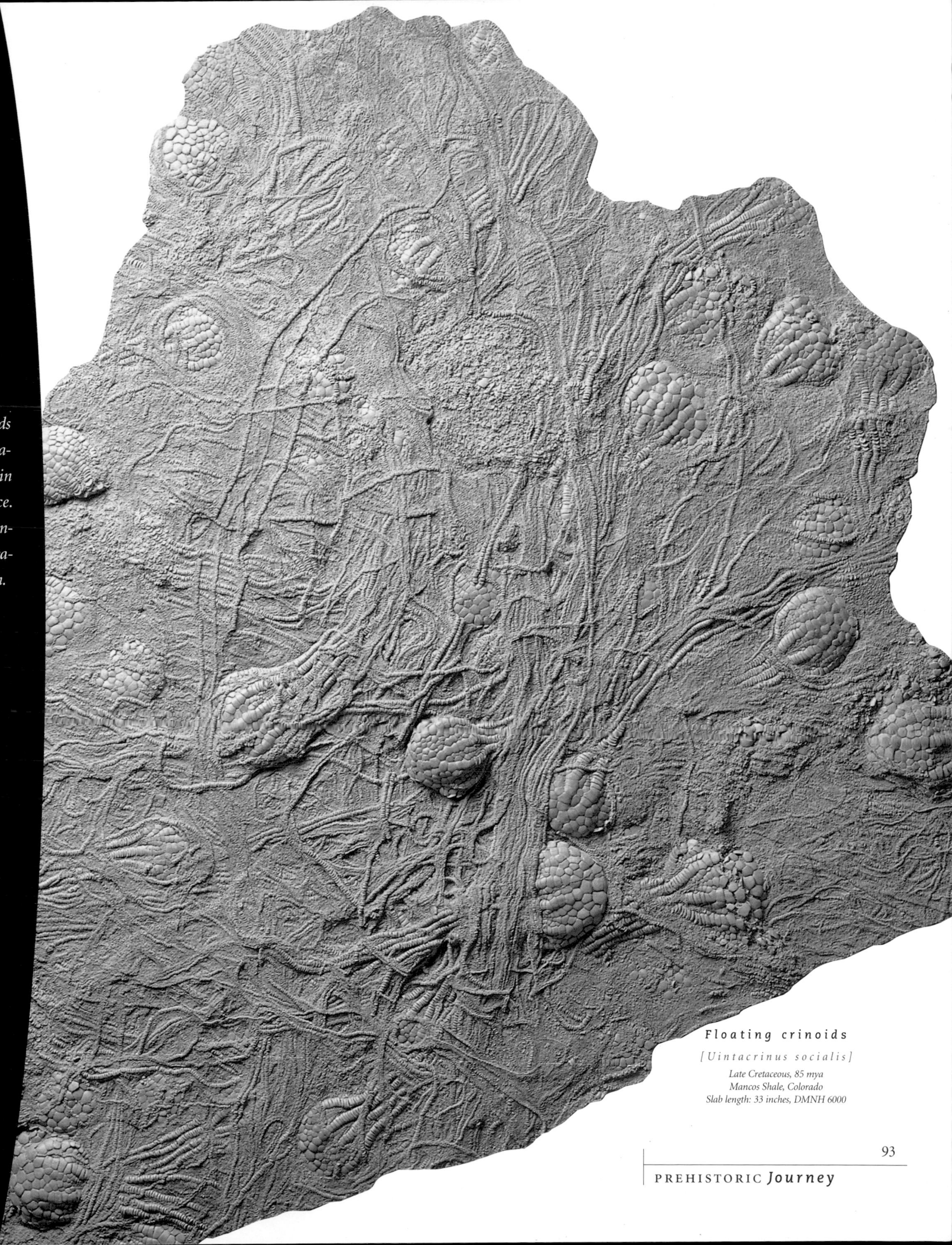

Floating crinoids

[Uintacrinus socialis]

Late Cretaceous, 85 mya
Mancos Shale, Colorado
Slab length: 33 inches, DMNH 6000

Denizens of the Cretaceous seas

The sea that covered the Western Interior of North America during the late Cretaceous period was populated by an array of giant marine creatures.

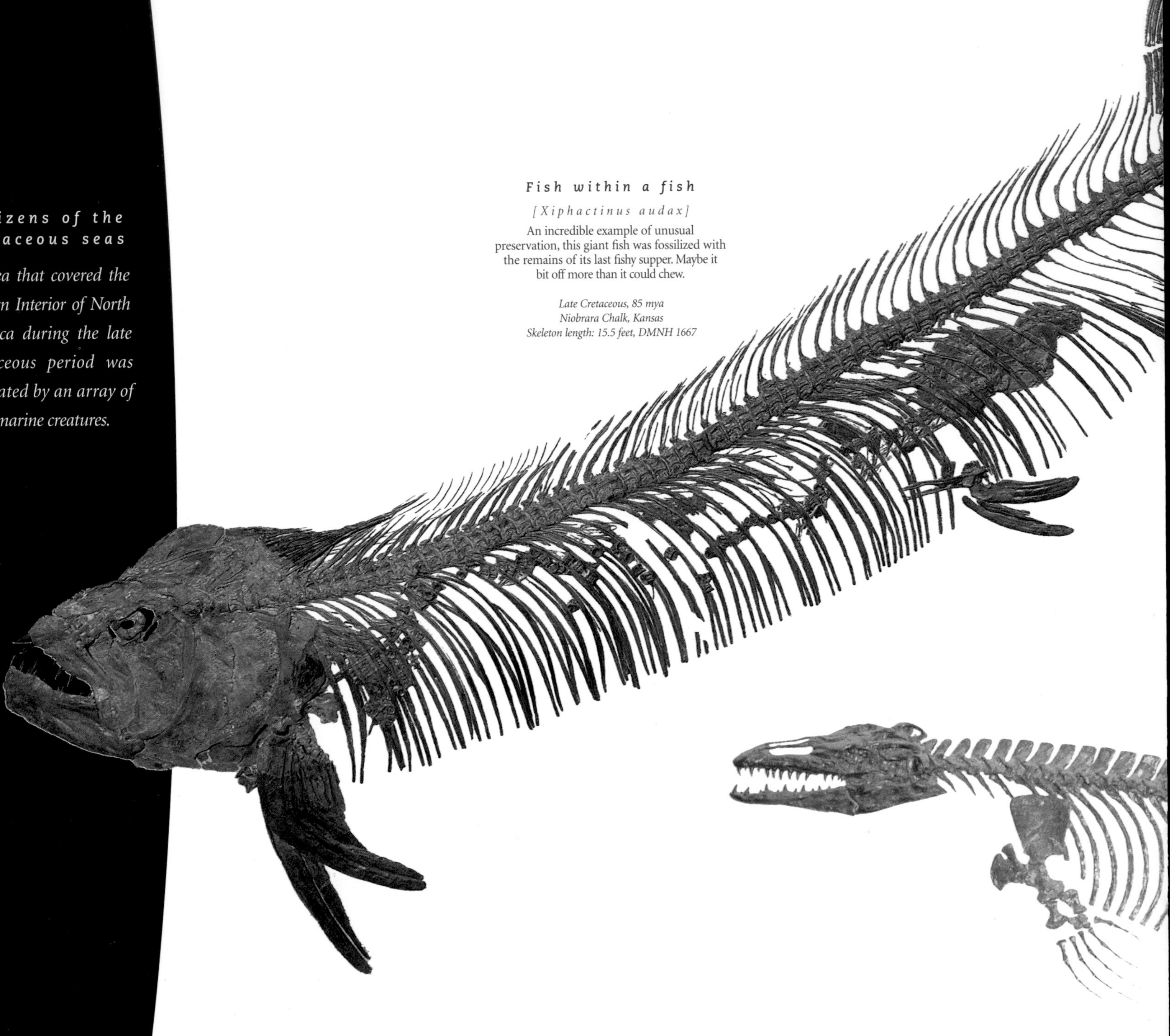

Fish within a fish

[Xiphactinus audax]

An incredible example of unusual preservation, this giant fish was fossilized with the remains of its last fishy supper. Maybe it bit off more than it could chew.

Late Cretaceous, 85 mya
Niobrara Chalk, Kansas
Skeleton length: 15.5 feet, DMNH 1667

Sea turtle

[Protostega gigas]

Cretaceous sea turtles maneuvered through the salty seawater of eastern Kansas much like their modern relatives paddle along the reefs of the Caribbean.

Late Cretaceous, 85 mya
Niobrara Chalk, Kansas
Skeleton length: 5 feet, DMNH 1663

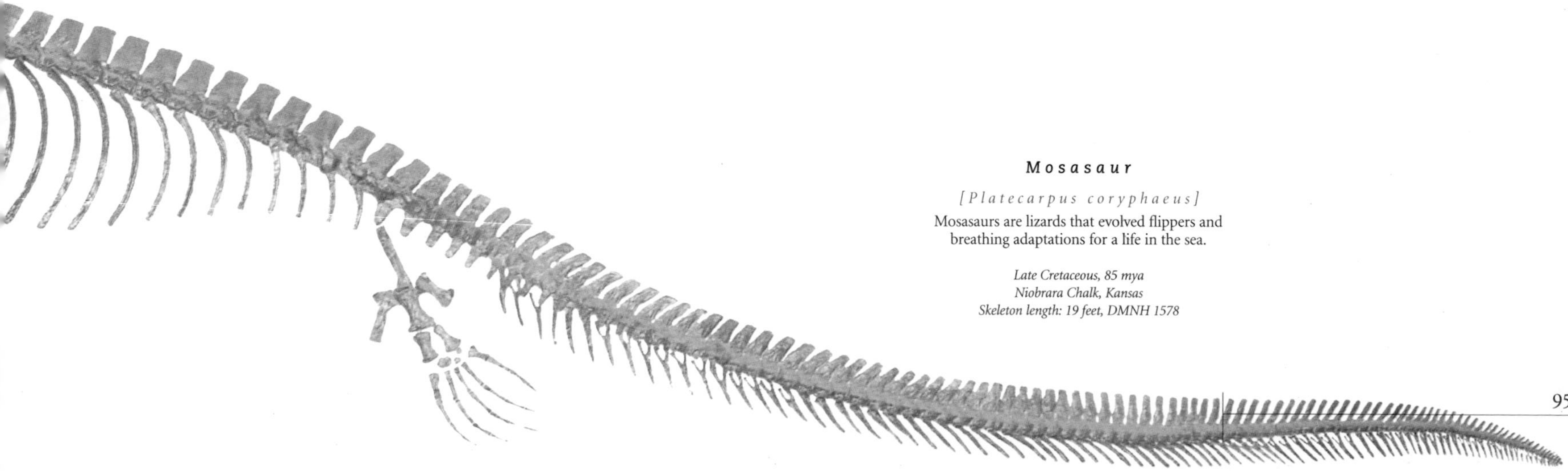

Mosasaur

[Platecarpus coryphaeus]

Mosasaurs are lizards that evolved flippers and breathing adaptations for a life in the sea.

Late Cretaceous, 85 mya
Niobrara Chalk, Kansas
Skeleton length: 19 feet, DMNH 1578

Ammonites, mollusks that are related to squid and octopus, had beautiful and elaborate shells. Like the living chambered nautilus and the diverse cephalopods of the Paleozoic, ammonites were advanced predatory animals that were well built for rapid swimming and foraging at various depths. Their shells were separated into many chambers by intricate walls. Peeling away the fossil shells of these animals reveals the junction where the wall contacted the shell, a complicated line known as a suture. The sutures are very useful for recognizing different species of ammonites. The ammonites of the seaway evolved rapidly and into a dazzling display of diversity. Some, like the baculites, grew straight and as long as 6 feet; others coiled in erratic patterns. Other Cretaceous ammonites grew to huge proportions, reaching diameters of 6 feet and weighing over a ton when alive. Many ammonite fossils are splendid because colors play off their pearly shells. Some of the finer specimens rival the colors of Australian fire opals. William Cobban, a U.S. Geological Survey paleontologist who has devoted his life to studying these creatures, has recognized almost 90 sequential assemblages of ammonites spanning 45 million years of the Cretaceous during which seas covered the center of the North American continent.

Straight ammonite

[Baculites cuneatus]

The delicate dendritic design on this baculite is its suture pattern. This pattern is visible only when the shell has been removed, exposing the edges of the elaborate walls that separate the mudstone-filled chambers of this animal. Baculites and other ammonites are distantly related to the modern chambered nautilus, which has a straight and simple suture pattern.

Late Cretaceous, 72.9 mya
Pierre Shale, South Dakota
Shell length: 9.8 inches, DMNH 7865

A scaphite couple

[Jeletzkytes nebrascensis]

Scaphites are a type of ammonite commonly found in marine rocks in the Western Interior of North America. Female (left) and male (right) scaphites had shells that were different in size and shape.

Late Cretaceous, 67.2 mya
Fox Hills Formation, South Dakota
Shell length: 4.9 inches, DMNH 6024
Shell length: 3.9 inches, DMNH 6023

Ammonites

These ammonites lived and died together in the shallow seas that covered Wyoming nearly 90 million years ago.

Ammonites

[Prionocyclus wyomingensis]

Late Cretaceous, 87 mya
Frontier Formation, Wyoming
Slab length: 16.5 inches, DMNH 6006

Bundle of baculites

Baculites were straight-shelled relatives of the coiled ammonites. They were common animals in the shallow seas that covered North America at the end of the Cretaceous. Their long pipe-like shells are often found in groups, suggesting that they either lived (and died) in schools or that they were accumulated by underwater currents after they had died.

Bundle of baculites

[Baculites compressus]

Late Cretaceous, 73 mya
Pierre Shale Formation, South Dakota
Specimen height: 13.8 inches, DMNH 6808

Ash layers from ancient volcanic eruptions provide a time frame for calibrating rates of ammonite evolution. Volcanoes in western Montana and Utah periodically rained volcanic ash into the seaway that covered the American West. The ash settled to the seafloor and was deposited as white layers in the otherwise dark gray seafloor mud. There are hundreds of ash layers in the resulting marine rocks. Each layer records the age of its level. Together, the ammonites and the ash layers make the Cretaceous marine rocks that are between 112 and 67 million years old one of the best dated evolutionary sequences in the world.

Back on land for the end of the Cretaceous

The coasts of the Cretaceous seaway were the habitat for a great variety of animals and plants. It is from the sediments there that we have some of our best records of late Cretaceous dinosaurs, mammals, and plants. Among the great coastal deposits of the period is that of Dinosaur Provincial Park in Alberta, Canada; in a tiny swath of badlands along the Red Deer River are preserved the remains of many dinosaurs. Much of the sediment that composes the rock at the park was deposited rapidly in tidal estuaries along the coast of the seaway. Dinosaurs that died here were rapidly buried and as a result are often preserved as whole skeletons. Many fossil dinosaur carcasses even preserve imprints of their tough scaly skin. Over 300 complete skeletons representing about 40 species of dinosaurs have been recovered from the park. A portion of the late Cretaceous, between 80 and 70 million years ago, represents the apex of dinosaur evolution. Worldwide, over 70 genera of dinosaurs are known from this time. These animals were quite different from the giant sauropods of the Jurassic. The Cretaceous dinosaurs were much smaller; many were the size of a human adult or even a child. The common types included the duck-billed hadrosaurs, the horned ceratopsians, the armored ankylosaurs, the sharp-toothed dromaeosaurs and troodontids, and the bone-headed pachycephalosaurs. Although the herbivorous dinosaurs were smaller than their Jurassic counterparts, some of the late Cretaceous carnivorous dinosaurs, such as the tyrannosaurids, were larger.

Heteromorph ammonite

[Didymoceras nebrascense]

Heteromorph ammonites are ones whose shells coiled in an apparently erratic manner. It is not clear how these animals swam; their lifestyle remains a mystery.

Late Cretaceous, 75.9 mya
Pierre Shale, South Dakota
Specimen length: 9.4 inches, DMNH 5900

Iridescent ammonite

[Sphenodiscus lenticularis]

The brilliant red iridescence of this ammonite was formed when fossilization deformed the microscopic structure of the shell.

Late Cretaceous, 68 mya
Fox Hills Formation, South Dakota
Shell diameter: 15.7 inches, DMNH 6009

Although the late Cretaceous saw a tremendous diversification of dinosaurs, mammals remained in the ecological background. Mammals had been present since the Triassic, but their diversity remained low. Most of the Cretaceous mammals were small insectivorous creatures. Four major groups were present: the monotremes of Australia; the marsupials, distant ancestors of Australia's hopping fauna; the multituberculates, rodent-sized beasts with odd teeth; and small placental mammals.

Our knowledge of the plant life of this part of the Cretaceous was recently enhanced by the discovery of a fossil site that was formed when volcanic ash rained down on a late Cretaceous landscape 71 million years ago. The site was accidentally discovered by Scott Wing, a Smithsonian Institution paleobotanist, in central Wyoming in 1991. The ash had been deposited quickly enough to preserve tender herbaceous plants in growth position. Delicate fronds of ferns, small wildflowers, tiny cycads, and complete palm plants are preserved where they grew in this "plant Pompeii." This unusually intact site provides a glimpse of a landscape that existed just before angiosperms rose to ecological dominance. Although flowering plants had been common on earth since the beginning of the Cretaceous, they had minor roles in many ecosystems until the very end of this period. The Wyoming site preserves a vast fern, cycad, and angiosperm prairie and catches the angiosperms in the act of taking over. It must have been vegetation like this that supported the diverse dinosaurian faunas of the late Cretaceous.

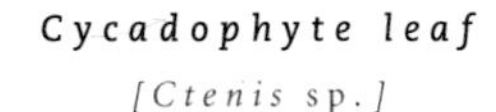

Cycadophyte leaf

[Ctenis sp.]

Cycads and their relatives were common components of late Cretaceous herbaceous vegetation. Very few of these plants survived into the Tertiary period.

Late Cretaceous, 70 mya
Almond Formation, Wyoming
Leaf width: 6 inches, DMNH 7905

But this vegetation and the dinosaurs of the Dinosaur Provincial Park were both replaced, a mere 2 to 3 million years before the end of the Cretaceous, by the dense angiosperm woodlands and the dinosaur fauna of the Hell Creek Formation. Though we are unsure of the causes of these changes, they are matched by other events elsewhere in the world at the same time. The majority of the tropical reefs in the world in the late Cretaceous were made not of corals but of cone-shaped mollusks known as rudists. Both rudist reefs and the giant inoceramid clams suffered major extinctions 2 to 3 million years before the Cretaceous period was terminated by the famous K-T boundary extinctions.

The Cretaceous-Tertiary boundary extinctions

No other event in earth's history has evoked as much speculation as the extinction of the dinosaurs. The end of the Cretaceous period, about 65 million years ago, is marked not only by the extinction of the dinosaurs but also by the demise of flying reptiles and most of the marine reptiles, all of the ammonites and baculites, many North American land plants and marsupials, and a great percentage of plankton in the world's oceans. Whatever killed the dinosaurs must have been pervasive enough to affect all these life forms.

The end of the Cretaceous is known as the Cretaceous-Tertiary, or K-T, boundary (*K* for *Kreide,* which is German for chalk), the literal physical boundary between rocks deposited in the Cretaceous period and younger ones deposited in the Tertiary period. Studies of the K-T boundary were greatly accelerated in 1980 when a group of University of California scientists led by Walter and Luis Alvarez discovered an unusually high abundance of the metal iridium at the K-T boundary in Italy, Denmark, and New Zealand. Because iridium is a rare substance on the earth's surface but common in meteorites, they proposed that this particular anomaly was caused by the impact of a 6-mile-wide asteroid. The headlines in the popular press blared: "An asteroid killed the dinosaurs!" This proposal landed on the unprepared and unreceptive ears of the scientific community. To paleontologists, it seemed like science according to the tabloid press.

But the theory could be tested: Was the iridium anomaly located precisely at the K-T boundary? Could it be found at many places on the globe? Were dinosaur fossils found above it? How were other organisms affected? Was there other physical evidence of an extraterrestrial impact? Where was the crater? These questions provoked a massive surge of K-T boundary research. Opposing the theory were scientists who felt that earthbound volcanism was a more likely source of the iridium. To support their claims they pointed to the Deccan Traps, a massive deposit of lava that flowed out of the earth in what is now western India in the late Cretaceous. Paleontologists were also reluctant to accept the new villain, but it became clear that many had not looked very closely at the time resolution of the fossil record.

Although the K-T boundary is still cause for heated discussion, several major discoveries in the 1980s support the asteroid-impact hypothesis. The iridium-bearing layer has now been found at more than 100 sites scattered around the world, and the boundary layer has been shown to contain shocked mineral grains of a type typically found in asteroid impact craters and nuclear test sites. In 1991, Alan Hildebrand, a Canadian geologist, rediscovered a giant buried-impact crater in late Cretaceous rocks on Mexico's Yucatán peninsula. The crater was originally found in 1981 but somehow escaped the eye of science until 1991. Clearly, an explosive impact had occurred, but was it the cause of the extinctions?

In the years since the asteroid theory was proposed, many paleontologists have undertaken detailed studies of fossils across the K-T boundary to see if the iridium-bearing layer occurs at the exact level of the extinctions. The most precise results can be achieved from microfossils because they are so small and abundant. A single teaspoon of sediment can yield thousands of specimens. Early results from studies of microscopic fossil pollen and spores in land deposits and fossil plankton in marine deposits showed an incredible coincidence of iridium enrichment and extinction. Robert Tschudy, a palynologist from the U. S. Geological Survey, discovered that fern spores represented almost all of the flora just above the iridium layer and suggested that the high abundance of ferns was due to their rapid colonization of the world after it had been devastated by the aftereffects of an asteroid impact. He recalled scientific studies of how plants had recolonized the Indonesian volcanic island of Krakatau in the years after it had erupted in 1883 and killed everything on it. Ferns were the first plants to colonize Krakatau, and for many years after the eruption, only ferns grew there.

Detailed studies of ammonite extinction by Peter Ward of the University of Washington and of land plants by Kirk Johnson of the Denver Museum of Natural History have confirmed that the extinctions were abrupt and coincident with the iridium layer. Many vertebrate paleontologists have been reluctant to accept an extraterrestrial executioner. And although no studies have shown that dinosaurs survived the K-T impact, many scientists still argue that dinosaurs were declining before the impact and suffered a less dramatic fate. The widespread nature of the biological catastrophe, the abrupt disappearance of the Hell Creek dinosaur fauna, and the ubiquitous presence of the iridium anomaly seem to be strong arguments for an extraterrestrial end to the age of reptiles. Whatever the cause of the K-T boundary extinctions, the effect was phenomenal. With the large dinosaurs removed, the stage was set for the age of the mammals and the rise of the modern world.

Living turtle and dead dinosaur

[*Neurankylus* and *Triceratops*]

The end of the Cretaceous saw the extinction of the dinosaurs. Members of other groups of reptiles, such as the snakes, lizards, and turtles, survived, and their descendants are alive today.

Lost cabin, wyoming

[50 Million Years Ago]

A troop of the lemur-like primate Notharctus *moves through the dense tropical forest canopy. A mother carrying her young infant leaps from a sycamore tree toward the delicate branches of a nearby legume where she plans to forage on the abundant pea-filled pods. A male follows and watches the apprehensive infant and the determined mother. Butterflies pause on the tangled vines of the climbing fern* Lygodium. *The forest is alive with the early morning ounds of birds and mammals. An opening in the branches allows a glimpse of a hippo-like* Coryphodon *crossing a muddy stream. Nearby, the dawn horse,* Hyracotherium, *is stalked through the thick forest underbrush by the agile carnivorous mammal* Protomus. *In the distance rise the newly formed Rocky Mountains, blanketed with a dense subtropical rainforest.*

chapter EIGHT

8

Origin of Modern Animals:

The Early Cenozoic

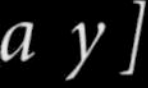

Majestic antelope and trophy bucks often prance through the canyons and valleys at Buck Spring, close to the hamlet of Lost Cabin in central Wyoming. A prickly pear and sagebrush prairie blankets the treeless lowlands at the base of the Bighorn Mountains, here and there revealing isolated patches of candy-striped badlands that are exposures of the 50-million-year-old Wind River Formation. This formation is a spectacular graveyard of ancient ecological communities from the time when the Rocky Mountains rose.

During the final stages of uplift in the Rocky Mountains, streams and rivers deposited gravel, sand, and mud in the many basins of the intermountain regions of the American West. Trillions of tons of mud and sand were eroded from highlands and carried by river systems into the Wind River Basin. Buck Spring is only a few miles south of the Bighorn Mountains, and nearby deposits of large boulders show that rocky streams once flowed off those mountains and into the basin. Fossils in these basin-filling sedimentary rocks provide a glimpse of ancient Rocky Mountain life 50 million years ago, about 15 million years after dinosaurs became extinct.

Serendipity often defines fossil discoveries, as it did for the Buck Spring Quarries. Paleontologists from the American Museum of Natural History, Princeton University, and the Carnegie Museum of Natural History had scoured these rocks for the isolated jaws and teeth of fossil mammals since the late 1800s. Few skeletons or skulls had been found. Weathering and winter freezing usually destroy the delicate fossils in these deposits before erosion exposes them. In 1984, Leonard Krishtalka and Richard Stucky, then of the Carnegie Museum, were prowling the badlands for fossils on what was supposed to be the last day of the summer field season. Stucky found a few scattered vertebrae lying on the surface of one gray mudstone hill. He decided to investigate, used his pick to lift up a clump of rock, and there lying on its side was the skull and front part of the skeleton of *Hyracotherium*, the dawn horse. As the excavation proceeded for the next month, tiny complete bones, jaws, and even a few skulls of mouse-sized animals were exposed. One specimen was the skull of an early Eocene primate, *Shoshonius*. This was only the second skull of its kind ever found! The first had been discovered in 1881.

The fossils at Buck Spring are found in small limestone lenses that are only a few feet thick, less than 15 feet wide, and lie a few hundred yards from the sandstone remnants of ancient meandering streams. These lenses, which were once small ponds on the ancient floodplain, contain rich bone beds that preserve more than 100 species of mammals and reptiles, including lizards, primates, carnivores, rodents, insectivores, marsupials, and several groups ancestral to modern plant-eating animals.

Evidence of the vegetation of the Buck Spring habitat comes from a nearby fossil plant site known as Dolus Hill. This site has yielded fossils of more than 70 species of broad-leafed plants. The large leaves with their oblong shape, smooth edges, and distinctive drip tips are strikingly similar to those found in today's tropical rainforests. Many of the fossil plants from Dolus Hill also have relatives in the modern subtropical forests of Central America. The diversity and physical characteristics of the tropical flora and fauna provide strong evidence that the early Eocene was one of the warmest times in earth's history.

Creodont skull

[Prototomus vulpeculus]

By 50 million years ago, a diverse number of flesh-eating mammals had evolved. The creodonts were voracious predators that fed on the dawn horse and other early ancestors of modern mammals. This skull was crushed and distorted by being buried and fossilized.

Early Eocene, 50 mya
Wind River Formation, Wyoming
Skull length: 6.3 inches, DMNH 6080

Rainforest sycamore leaf

[Macginitiea gracilis]

The lobed leaves of ancestral sycamores are some of the more common leaf fossils from the Eocene rainforest of Wyoming.

Early Eocene, 50 mya
Wind River Formation, Wyoming
Slab width: 1.3 feet, DMNH 7898

chapter EIGHT

Origin of Modern Mammals: The Early Cenozoic

The Eocene World

[50 MILLION YEARS AGO]

A greenhouse atmosphere creates a warm earth with no polar ice caps. The modern continents are taking shape.

The tropical Rockies: adaptations to a warm world

The extremely well-preserved late Cretaceous and early Tertiary fossil record of the Western Interior is a result of the formation of the modern Rocky Mountains during what is called the Laramide Orogeny. Beginning about 70 million years ago, giant slabs of the earth's crust were driven upward along steep faults, creating a series of mountains and basins that we know as the Rocky Mountain region. The uplifted areas were eroded by rain. Rivers carried mud and debris down from the mountains into the basins below. Layer upon layer of sediment accumulated in the low-lying basins, creating deposits thousands of feet thick. The basins sank slowly because of both the weight of the sediment and the movement of adjacent mountain ranges. As the basins subsided, more sediment was deposited. Tremendous thicknesses of sediment accumulated in relatively short periods of time. In some places in northern Wyoming, layers of sand and gravel 2 miles thick were deposited in less than 4 million years.

The uplift of the mountain ranges was mostly complete by 50 million years ago. It was followed by a series of explosive volcanic eruptions centered in the area that is now Yellowstone National Park. These eruptions spewed volcanic ash, intermittently blanketing much of central and western North America for many millions of years. The river-borne sediments and airborne volcanic ash entombed the plants and animals that lived in and near these basins. The ever-subsiding basins buried and preserved a near-continuous sample of animals and plants that inhabited the region.

Sixty-five million years ago, at the start of the Tertiary period, the continents were nearly in the positions that they are today. North America was a bit farther north and rotated slightly clockwise from its present position. It was connected to Europe by a land bridge from Greenland to Scandinavia and to Asia by a span from Alaska to eastern Siberia. Africa and South America were slightly closer to one another than they are today. Australia was connected to Antarctica, and India was an island that was moving rapidly north across the Indian Ocean on a collision course with Asia. Because of the positions of the continents in the early Cenozoic, the circulation of the oceans was more restricted than today.

The development of mountain ranges and other physical changes were accompanied by climatic changes. Global warming occurred due to high levels of carbon dioxide in the atmosphere as well as the circulation of the oceans. The high level of carbon dioxide resulted in both higher temperatures and a less seasonal climate than exist today. Polar ice was absent and rainfall was high, even in the interior of continents where today deserts are now widespread. Forests grew as far north and south as there was land. The early Cenozoic world was tropical across regions that today are temperate. In this tropical milieu, animals that survived the event that destroyed the dinosaurs diversified and evolved into the ancestors of our modern mammal, bird, and reptile families. It was a world of ecological and evolutionary opportunity of revolutionary proportions.

Paleocene mammal skull

[*Baioconodon denverensis*]

The mammals that lived immediately after the extinction of the dinosaurs were all very generalized. They were awkward runners and probably fed on plants and insects.

Early Paleocene, 65 mya
Denver Formation, Colorado
Skull length: 3.4 inches, DMNH 2500

Palm frond

Fossil palm leaves from the Green River Formation in Wyoming are some of the most convincing evidence that the climate of the Eocene was much warmer and wetter than today.

Palm frond

[Palmacites sp.]

Middle Eocene, 50 mya
Green River Formation, Wyoming
Leaf length: 5.4 feet, DMNH 6807

Primate skeleton

Primates were very common when North America was covered by tropical forests. In some fossil sites more than a dozen different kinds of primates are found together.

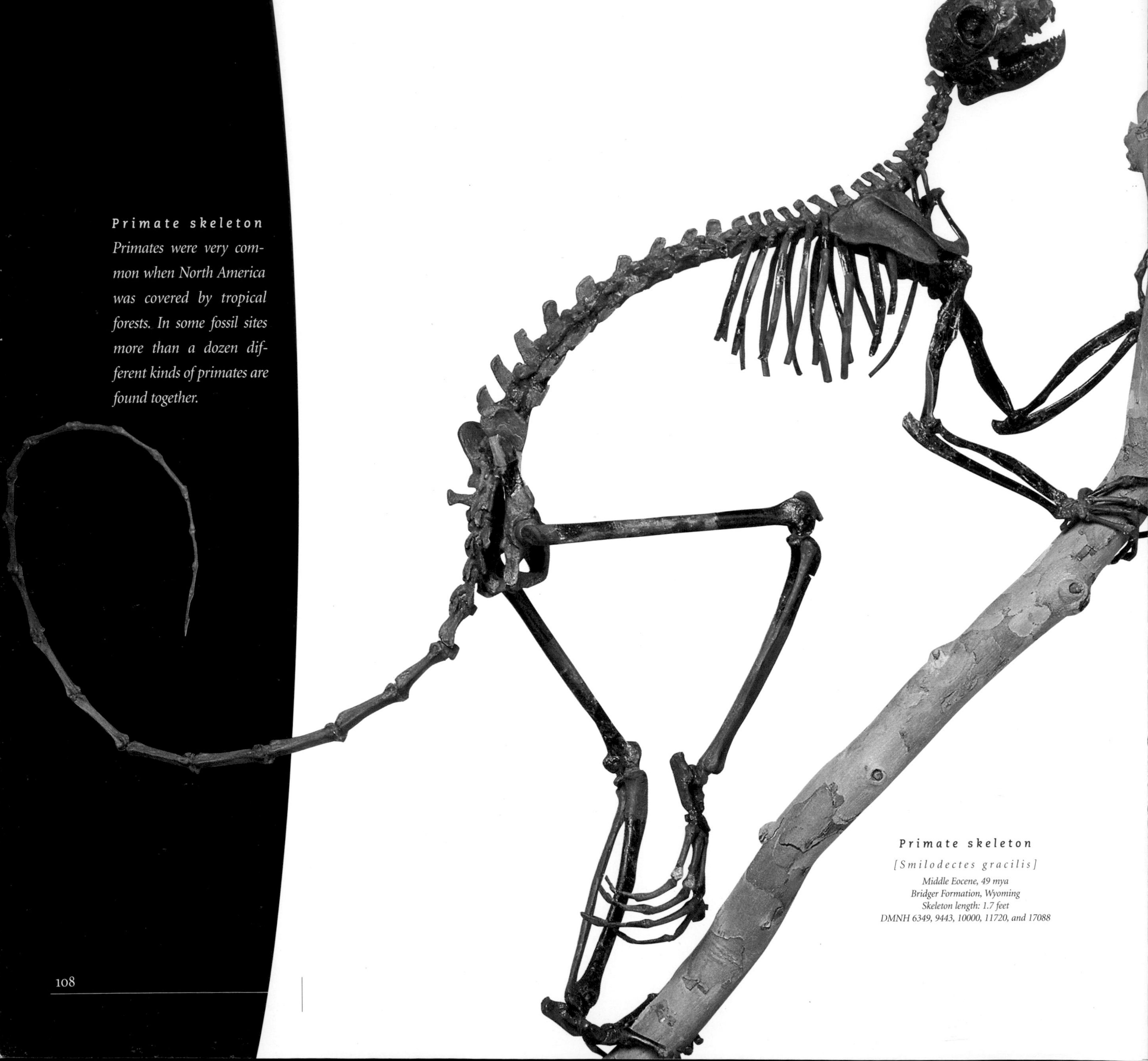

Primate skeleton

[*Smilodectes gracilis*]

Middle Eocene, 49 mya
Bridger Formation, Wyoming
Skeleton length: 1.7 feet
DMNH 6349, 9443, 10000, 11720, and 17088

The world had changed dramatically from the time of the demise of the dinosaurs. The extinction at the end of the Cretaceous left the world with simpler ecosystems; the plant- and flesh-eating specialist dinosaurs and many plants were gone. Some paleontologists speculate that vegetation became denser, providing more places for animals to hide and creating a sullen darkness on the forest floor at midday. Subtropical forests with few kinds of trees dominated the landscape of the North American midcontinent. Many plants responded to the closing of the canopy by increasing the size of seeds to store more energy for early growth in the low light levels of the forest floor.

These features made the post-dinosaur world, or early Tertiary, a place where the animals that survived the extinction could flourish. Before the extinction of the dinosaurs, only a few dozen mammal families existed worldwide. They were all small, ranging in size from a mouse to a house cat; they were bound to secretive habitats in the underbrush where the ability to maneuver through dense vegetation was a necessary skill for small mammals when dinosaurs towered overhead. Many were generalists, feeding on whatever food resources they came across. Soon after the extinction of the dinosaurs, mammals underwent one of the most spectacular rises in diversity of any terrestrial group in earth's history. Only a few million years after the extinction, mammal diversity rose from a few dozen families to nearly 120 families, comparable to the number that exists today. Although each continent evolved its own fauna, intermittent migrations served to mix and enrich these faunas, especially those of North America, Europe, and Asia.

The largest mammals in the Paleocene (early Tertiary), such as *Periptychus* and *Baioconodon,* were the size of cocker spaniels and were able to feed on leaves or browse from brushes and trees. The smallest, insectivores and tiny marsupials, had teeth for feeding almost exclusively on insects. Specialized teeth among the mammals—projecting front incisors, crushing and slicing premolars, and mortar-and-pestle-like molars—suggest that some had restricted diets and fed on the gum from trees or on fruits and seeds. Some groups became more at

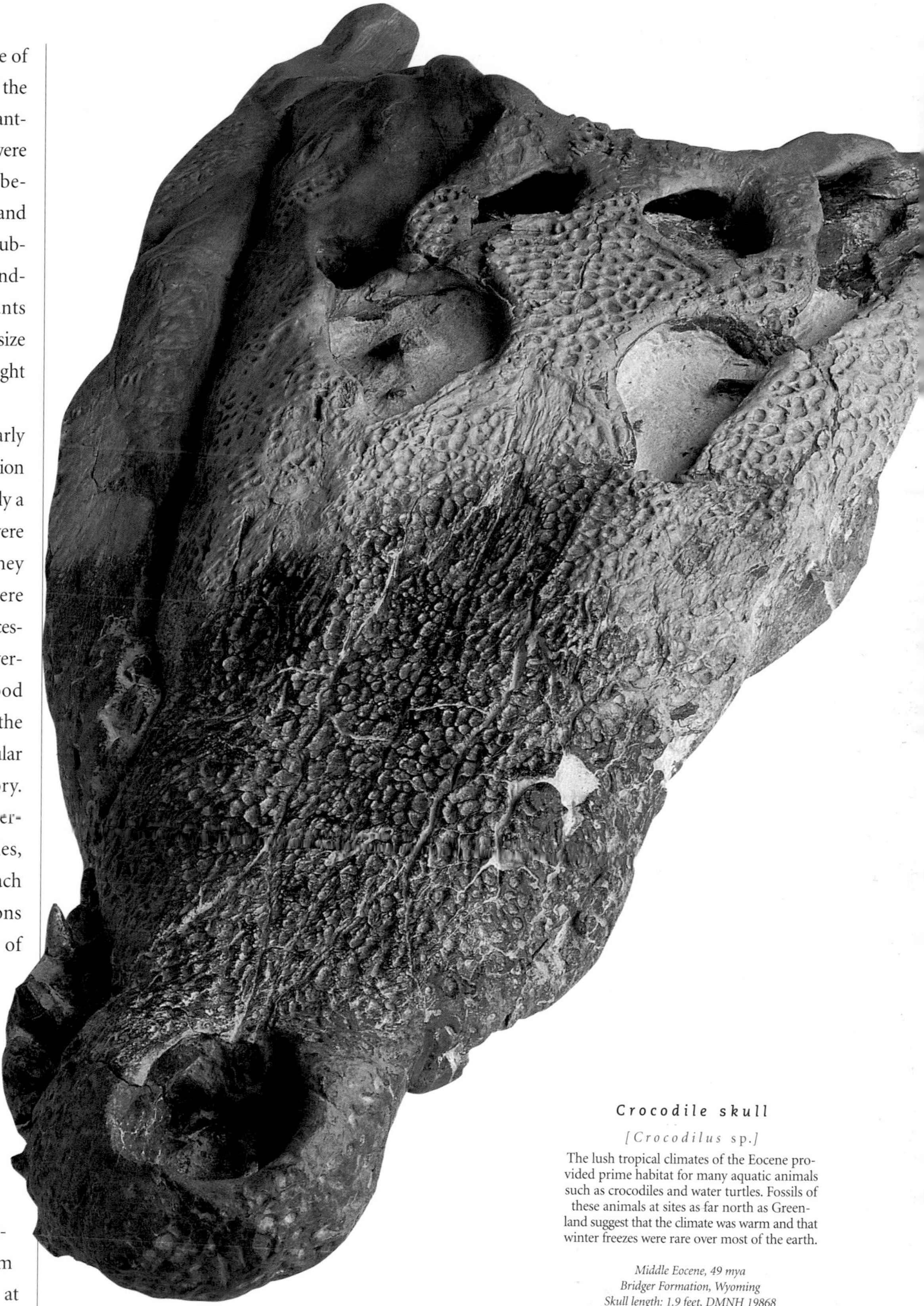

Crocodile skull

[Crocodilus sp.]

The lush tropical climates of the Eocene provided prime habitat for many aquatic animals such as crocodiles and water turtles. Fossils of these animals at sites as far north as Greenland suggest that the climate was warm and that winter freezes were rare over most of the earth.

Middle Eocene, 49 mya
Bridger Formation, Wyoming
Skull length: 1.9 feet, DMNH 19868

home in the upper reaches of the forest canopy, having the ability to glide through open spaces to their next haunt or to quickly move from branch to branch, using their agile forearms and hind limbs.

During the early Eocene epoch, the epoch following the Paleocene, the highest temperatures during the entire Tertiary were achieved. Palm trees grew in southern Canada and crocodiles lived above the Arctic Circle. Forests flourished in the most polar areas of land in North America and Antarctica. The climatic gradient of cold poles and hot tropics that characterizes the modern world did not exist. Instead there were mild polar regions and tropical regions that extended into the middle of North America, Europe, and Asia.

This expansion of warm climatic belts prompted a dramatic increase in plant types. Many genera of modern plants first appeared at this time. The lowlands of the Western Interior of North America supported a diverse subtropical to tropical rainforest. In the mountainous and high-latitude areas, a seasonal deciduous broadleaf vegetation developed. Today, remnants of these forests survive in the eastern United States and in eastern Asia.

By 50 million years ago, most of the modern major groups, or orders, of mammals had appeared. At the peak of tropical forestation, more types of mammals lived on the North American tropical landscape than at any time since. Among these were the first primates who were abundant in North America, Europe, and Asia. In North America, nearly a dozen species of primates inhabited the same community. *Notharctus* and *Smilodectes,* both the size of a house cat, resembled modern-day lemurs of Madagascar. Mouse-sized *Shoshonius, Omomys,* and *Absarokius* resembled the tarsiers of southeast Asia and mouse lemurs of Madagascar. These early primates were well adapted to living in the upper forest canopy because of their grasping hands and feet and the great mobility of their arms and legs. They had large eyes in the front of the head, which enabled them to see in three dimensions and in the low forest light, and a brain relatively larger than that of other mammals living at that time. *Shoshonius* had a very specialized blood circulatory system for delivering oxygen to the brain. This system is much like that in modern monkeys, apes, and humans. Its presence suggests that the first rung on the human evolutionary ladder had begun 50 million years ago.

Primate skull

[Smilodectes gracilis]

Large, forward-facing eye sockets on the skull of this Eocene primate suggest that it had binocular vision and accurate depth perception, traits needed for an arboreal lifestyle.

Middle Eocene, 49 mya
Bridger Formation, Wyoming
Skull length: 2.2 inches, DMNH 10000

Bat skeleton

[Palaeochiropteryx tupaiodon]

Within 15 million years after the extinction of the dinosaurs, mammals had evolved into a dazzling array of body types including flying forms such as the first bats.

Middle Eocene, 49 mya
Messel Formation, Germany
Skeleton length: 3 inches, DMNH 2728

Teeming with life, the tropical forests were home to many carnivorous mammals and primitive, plant-eating mammals, such as the rodents and *Hyopsodus,* a weasel-shaped planteater. The first modern plant-eating hoofed, or ungulate, mammals appeared 55 million years ago. They can be divided into two groups: the even-toed ungulates, or artiodactyls, and the odd-toed ungulates, or perissodactyls. Artiodactyls were very rare during the early stages of their history, but by 45 million years ago they became more common than the perissodactyls. Today the artiodactyls are the most abundant ungulates; this group includes antelopes, giraffes, sheep, pigs, hippos, and deer. During the Eocene, perissodactyls were very diverse and were among the most common large and medium-size mammals. Today the perissodactyls are less diverse than the artiodactyls and include horses, rhinos, and tapirs. One of the earliest perissodactyls was *Hyracotherium,* the dawn horse, or eohippus. The front and hind limbs of many of these early mammals were more advanced than those of mammals that lived earlier, suggesting that natural selection favored creatures that could run more quickly to avoid predators.

Panther-like, raccoon-like, and weasel-like predators stalked Eocene prey, but several predators stand out.

Fish death layer

Knightia is the most abundant fish of the Green River lakes. Some fossil beds preserve literally millions of individuals that apparently died at one time as the chemistry of their lake changed.

Fish death layer

[*Knightia eocaena*]

Middle Eocene, 50 mya
Green River Formation, Wyoming
Average fish length: 3.1 inches
DMNH 15279

Garfish

[Lepisosteus simplex]

This garfish is remarkably similar to those that live today in the Mississippi River drainage. They are living fossils.

Middle Eocene, 50 mya
Green River Formation, Wyoming
Skeleton length: 3.7 feet, DMNH 1649

Diatryma was a 6-foot-tall carnivorous bird that was closely related to the chicken family, and *Machaeroides* was a cat-sized, saber-toothed creodont, a now extinct group of carnivorous mammals.

The burst of mammalian evolution also resulted in some of the most specialized orders of modern mammals: whales and bats. The first whales, fully marine mammals, were descendants of a terrestrial group of mammals, the mesonychids, who were closely related to the ungulates but were carnivorous. During the Eocene, the mesonychids were among the most agile and swiftest of meat-eating mammals as well as the largest. Mesonychids are thought to be related to whales because of the similarity of their teeth and the detailed anatomy of the ear region and brain. The protocetes, or earliest whales from the Eocene, have simple three-cusped teeth like those in the mesonychids. If it were not for the reduced nature of the limbs, it would be very difficult to distinguish the protocetes from mesonychids. It was not until the 1990s that we began to understand the evolution of the whales from this fully terrestrial group. Recently, Philip Gingerich at the University of Michigan discovered tiny 3-foot-long hind legs on the skeletons of *Basilosaurus,* a 50-foot-long Eocene whale from Egypt. Although these whales had been known for over a century from sites in Egypt and North America, the legs were unnoticed, perhaps because they are very small and were unexpected on a whale. Other recent finds from the Eocene in Pakistan have generated a sequence from amphibious proto-whale types to fully marine, large-bodied whales. The missing links between land-living mesonychids and the water-dwelling whales have now been discovered.

Bats, mammals with the ability to fly, first appeared in the forests of the Eocene. Although fossil teeth that look like those of bats have been found in older rocks, the first skeleton of a bat with wings dates from about 51 million years ago. This exquisitely preserved skeleton comes from an ancient lake deposit in Wyoming now known as the Green River Formation. At one site in Germany, Eocene bats have been found in oil shales that still preserve some of the fine details of their internal anatomy and the remnants of their last meals.

Some of the basins that formed during the Laramide Orogeny had no external water drainage. In the rainy world of the Eocene, these basins sometimes filled with water and formed huge lakes. The most famous of these basin-filling lakes are known collectively as the Green River lakes: Lake Uinta, Lake Gosiute, and Fossil Lake. These Great Lakes of the Eocene covered vast areas of southwestern Wyoming, northwestern Colorado, and northeastern Utah between 52 and 45 million years ago. Rich in organic matter from the resident algae and freshwater plankton, the oil shales of the Green River deposits may yield a future energy resource.

The fossils in the lake shales provide a detailed picture of the life forms that lived in the lakes and near their shores. The most common fossils are fishes, plants, and insects. Fish are extremely abundant in Fossil Lake and Lake Gosiute. *Knightia,* a herring, and the perch-like *Priscacara* were among the most common fish in Fossil Lake. Millions of *Knightia* fossils have been excavated from quarries near Kemmerer, Wyoming. The large predatory fishes of the Green River lakes include the bowfin and garpike that today live in the Mississippi drainage.

Long-horned borer beetle

[Cerambycidae]

Although color is rarely fossilized, the wing covers of this 48-million-year-old, long-horned borer beetle still preserve the patterns that decorated them so long ago.

Middle Eocene, 48 mya
Green River Formation, Colorado
Beetle width: 1.2 inches, DMNH 8393

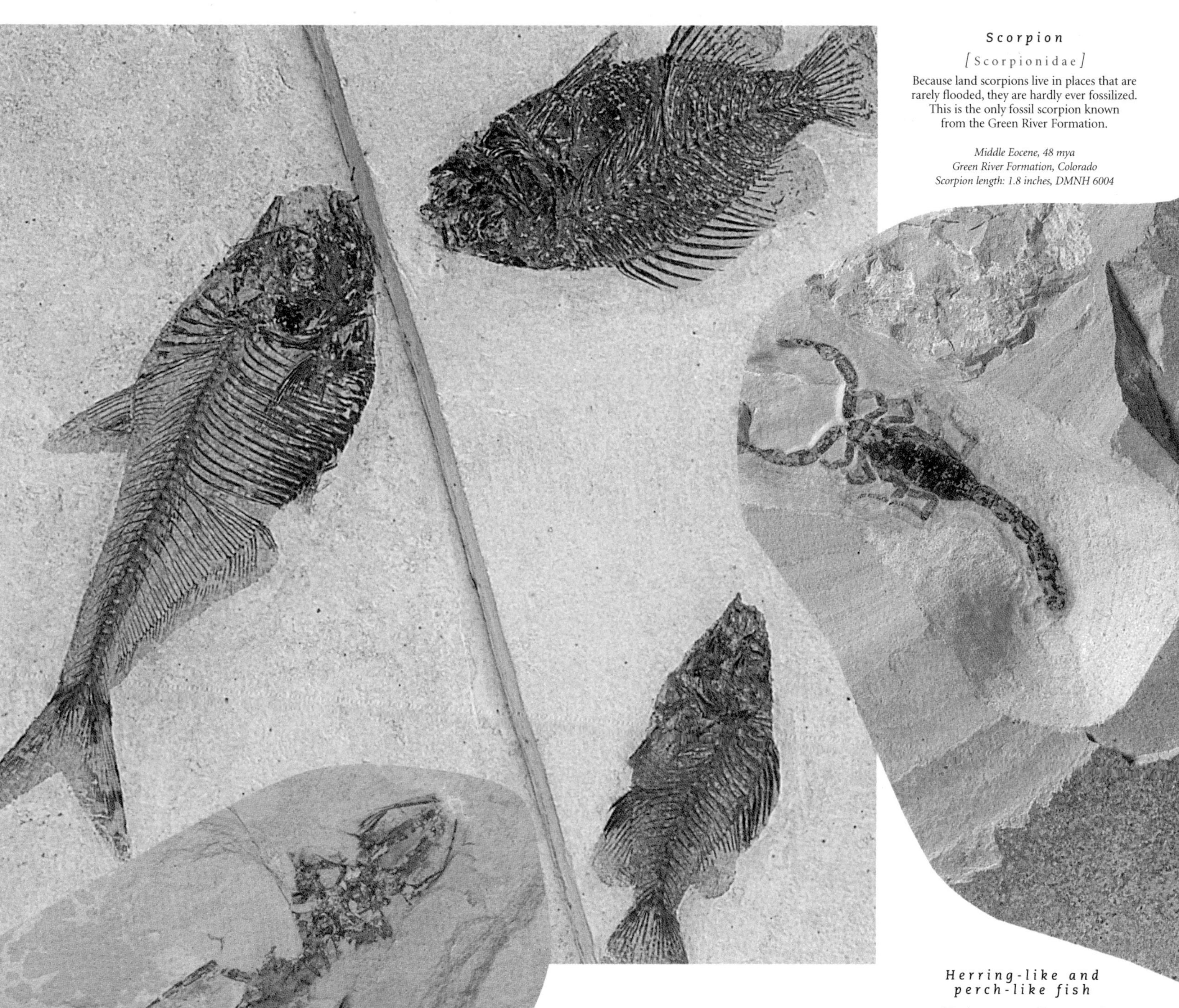

Scorpion

[Scorpionidae]

Because land scorpions live in places that are rarely flooded, they are hardly ever fossilized. This is the only fossil scorpion known from the Green River Formation.

Middle Eocene, 48 mya
Green River Formation, Colorado
Scorpion length: 1.8 inches, DMNH 6004

Frog

[Anura sp.]

Fossil frogs are most frequently found in the depths of ancient lakes.

Middle Eocene, 48 mya
Green River Formation, Colorado
Skeleton length: 1.7 inches, DMNH 15013

Herring-like and perch-like fish

[Priscacara liops and Diplomystus dentatus]

The Green River lakes occupied vast areas in the region of Utah, Wyoming, and Colorado. The lakes supported abundant fish and the fossil sites in these areas have produced millions of skeletons of perch, suckers, and other fish.

Middle Eocene, 50 mya
Green River Formation, Wyoming
Slab length: 1.1 feet, DMNH 1527

Plant and insect fossils are found at a number of Green River sites. The Green River flora is composed of plants that indicate the increasing importance of a dry season. Growing alongside spectacular large palms were over a hundred other species, such as willow, poplar, and sumac, that still survive in the American West. Hundreds of species of fossil insects from the Green River Formation include flies, ants, beetles, crickets, mantises, moths, butterflies, stick bugs, and dragonflies. Sites along the shores of the old lakes produce fossil crayfish and freshwater snails. The Green River Formation is also famous for its exquisite whole skeletons of snakes, lizards, frogs, turtles, crocodiles, mammals, and birds.

Sumac leaf

[Rhus nigricans]

Comparison of the vein patterns on this ancient sumac leaf with those of its modern relatives allows it to be identified and classified.

Middle Eocene, 48 mya
Green River Formation, Utah
Leaf lenght: 2.8 inches, DMNH 6948

The tropical world turns temperate: adaptations to a cooler and drier world

About 50 million years ago, the climate began to change and with it the character of the flora and fauna. The climate began to become cooler and drier. This change may have been caused by a decrease in atmospheric carbon dioxide and a greater degree of oceanic circulation between equatorial and polar regions as well as by the uplift of the Rockies in the central part of North America. The tropical broad-leafed evergreen forests started to give way to more seasonal and open forests. In practically all major groups of mammals, the size of the largest individuals increased.

A few primitive mammal groups were still present, although in greatly reduced numbers. One of these, the most bizarre of any North American mammal, was the uintathere, which had the body of a hippopotamus, three pairs of knobby horns across its face, and large saber-like canines. The odd-toed ungulates, the perissodactyls, were very well represented by a number of species. These included the titanotheres, rhinoceroses, tapirs, and a descendant of *Hyracotherium,* the horse *Epihippus*, which lived between 47 and 42 million years ago. The primates were in decline in North America presumably because of the breakup of the forest habitats; they eventually died out about 36 million

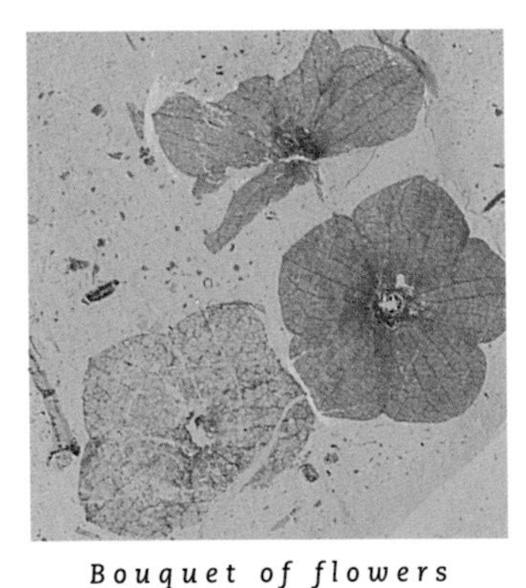

Bouquet of flowers

[Florissantia quilchenensis]

Fossil flowers are generally rare and a fossil bouquet is even rarer. These flowers grew on a tree related to the modern boxwood.

Middle Eocene, 48 mya
Klondike Formation, Washington
Slab length: 6.9 inches, DMNH 5098

Sabertoothed cat

[Dinictis squalidens]

At least four different lineages of predaceous mammals evolved a sabertooth adaptation.

Late Eocene, 35 mya
White River Formation, South Dakota
Skull length: 7.1 inches, DMNH 79

Uintathere

[Eobasileus cornutum]

With six horns, gigantic bodies, and saber-like canines, the uintatheres are among the most unusual of all mammals. They lived for only a few million years in North America and their fossils are quite rare.

Middle Eocene, 47 mya
Washakie Formation, Colorado
Skull length: 3 feet, DMNH 495-6

years ago. But the even-toed ungulates, the artiodactyls, began to increase in diversity and ecological importance. By about 47 million years ago, the first of a long line of successful artiodactyls, the oreodonts, appeared, as did relatives of the camel and animals resembling pigs. Rodents, which were also abundant during the early Tertiary, varied in size from that of a very small mouse to that of a badger and were much like modern squirrels, rats, and mice in their habits.

The Florissant Formation of south-central Colorado is another lake deposit that compares with the Green River in the preservation of life. It contains a flora and insect fauna and a handful of mammals, birds, and fishes that date from the late Eocene, 35 million years ago. The plants, which were very abundant, began to take on a more temperate character. The most abundant plant was *Fagopsis,* an extinct relative of the oak. Among the most interesting plants at Florissant were the world's first roses and the giant sequoias, whose trunk bases are often more than 12 feet in diameter. Other plants include a variety of conifers, such as pine and cedar, and deciduous angiosperms, such as hickory, sumac, willow, and poplar.

Perhaps the most spectacular badland fossil beds in North America are those of the White River Formation, which forms a veneer of sediment in the Great Plains from South Dakota south into Colorado and east into Nebraska. Badlands National Park in South Dakota was created to preserve the abrupt

The big stump

[Sequoia sp.]

The big stump at Florissant Fossil Beds National Monument is dramatic testimony to the reality of changing climates and ecosystems. Thirty-four million years ago, Colorado's mountains were covered with patches of giant redwoods. These trees now live only in California and Oregon.

The Pinnacles

[Western South Dakota]

The Pinnacles area at Badlands National Park in South Dakota is composed of weathering Eocene and Oligocene sedimentary rocks that contain superb skeletons of the animals that wandered North America as the climate cooled and landscapes began to change.

Creodont hunters

[HYAENODON AND MERYCOIDODON]

Powerful jaws and stealthful habits made Creodont one of the more feared predators of the late Eocene. Here a Creodont dispatches a hapless oreodont.

Creodont and oreodont

[Hyaenodon crucians and *Merycoidodon culbertsoni]*

Creodonts had long legs for running down prey and sharp, slicing teeth for killing and eating their victims.

Early Oligocene, 32 mya
White River Formation, Colorado

Merycoidodon skeleton length: 2.3 feet. DMNH 84/54

Hyaenodon skeleton length: 3.7 feet, DMNH 1604

pinnacles and cathedral-like badland outcrops. The White River Formation has produced some of the most complete fossils of mammals. Entombed between 37 and 28 million years ago, thousands of skulls and skeletons and tens of thousands of jaws and bones of fossil mammals have been discovered after a century of paleontological exploration. These are some of the world's richest fossil mammal beds. In 1994, a crew of six from the Denver Museum of Natural History recovered nearly 1,000 jaws from 1 square mile of outcrop in one day!

Most abundant among the mammals of this time were the oreodonts, odd-shaped artiodactyls, which resembled a cross between a pig and sheep. Forebears of the camel and antelope were also well represented. But towering over all the other mammals were the titanotheres, or thunder beasts, which had two horns on the tip of their snouts. As large as modern rhinoceroses, these browsers fed on the leaves and twigs of the forest. They became extinct about 34 million years ago. True rhinoceroses were abundant, as were the primitive three-toed horses called *Mesohippus.* The carnivores had become modern in appearance with cat-like, bear-like, and weasel-like forms. The first dogs appeared also. The most voracious carnivorous mammals then, however, were the primitive creodonts such as *Hyaenodon,* which had a battery of sharp scissor-like molars for slicing prey into bits.

Squirrel relatives and mice were also abundant. The first beavers appeared, but they were fully terrestrial in habit, probably burrowing like the other rodents of the day. A few very strange and surprising mammals include *Xenocranium,* a mole-like predator that fed on ground-living insects and worms. It had a shovel-like nose and powerful forelimbs for digging. *Patriomanis* was a North American scaly anteater related to the pangolins of Asia and Africa.

As the savannas advanced in North America, the animal communities took on a new character. Some extremely rich fossil quarries preserve the bones of many individuals of one species, such as the rhinoceros *Trigonias,* suggesting that these animals lived in herds. The males and females of some of these herbivores were different in size, and within some groups the males had horns. The plant-eating animals appear to have been able to run more swiftly than their ancestors. Animals with strong front limbs for burrowing also were more abundant. All of this evidence suggests that the more open habitats were key for the evolution of more complex social behavior and visual recognition of individuals of one's species. At the same time, animals were becoming larger and swifter or they simply began to burrow to avoid predators in the newly opened landscape.

In the equatorial continents of Africa, Asia, and South America, tropical rainforests persisted and have continued until modern times. During the late Eocene as the climate became cooler and drier, the tropical ecosystems became compressed into a narrower belt around the equator. The first anthropoid primates—ancestors of monkeys, apes, and humans—appeared in these tropical regions in Egypt about 35 million years ago. Some rare East Asian fossils that are about 45 million years old may be the first evidence of the ancestral anthropoid stock. Australia was home to the marsupials that evolved in similar fashion to the placental groups of the northern continents, and South America had a unique mixture of marsupial and placental mammals. Africa was home to a number of modern groups, including elephant ancestors. By 35 million years ago, the world was fragmenting into distinct tropical and temperate ecosystems. Mammals were beginning to show complex social interaction, one of the hallmarks of the modern world.

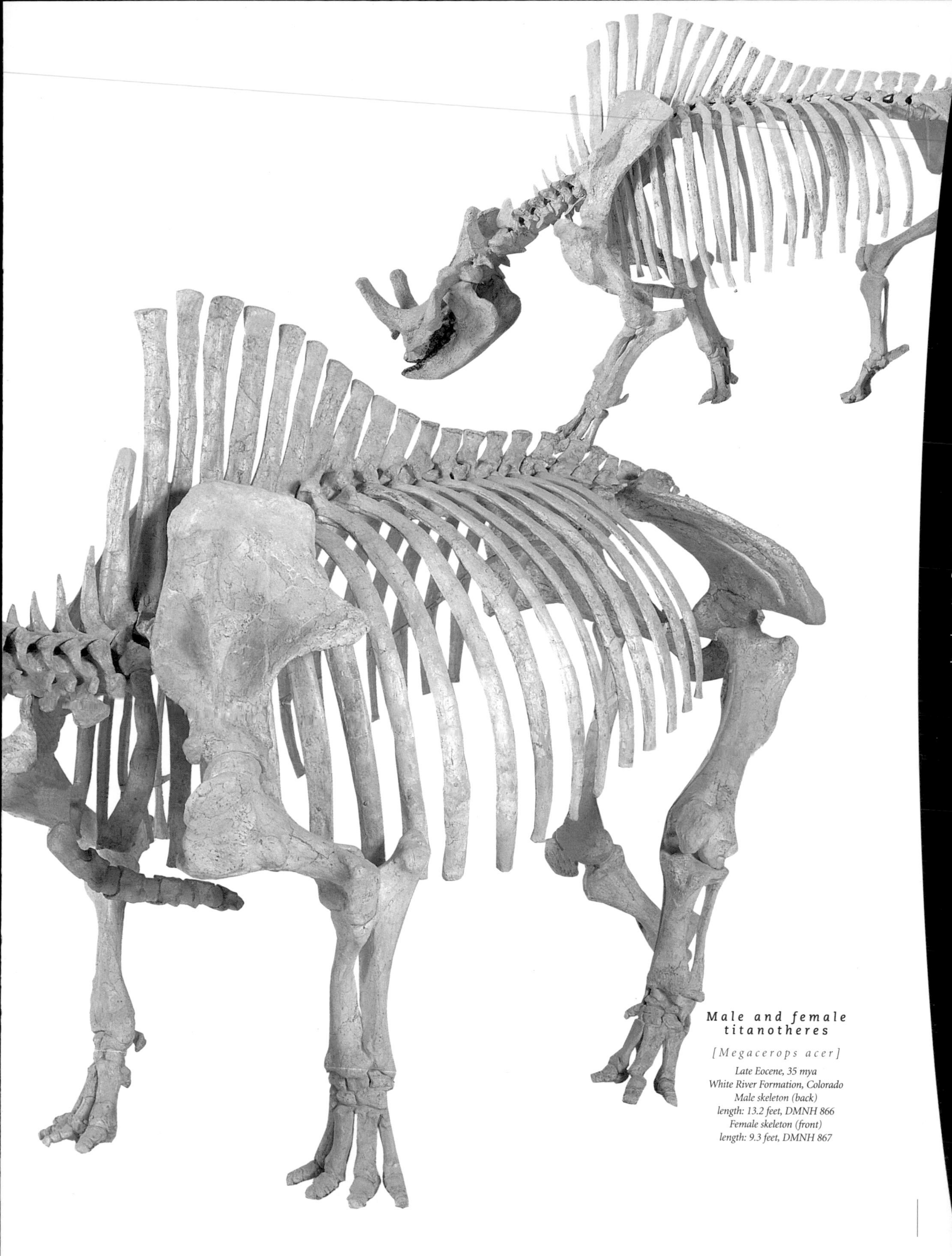

Male and female titanotheres

[Megacerops acer]

Late Eocene, 35 mya
White River Formation, Colorado
Male skeleton (back)
length: 13.2 feet, DMNH 866
Female skeleton (front)
length: 9.3 feet, DMNH 867

Male and female titanotheres

Cooling and drying of the climate at the end of the Eocene epoch caused the landscape to become less forested. As mammals adapted to the changing landscape, they grew larger, formed herds, and became fast runners. In the opening landscape, visual recognition became more important, causing the evolution of features such as differences in body size between males and females and elaborate head ornamentation.

Agate springs, Nebraska

[20 Million Years Ago]

A mid-afternoon thunder-storm looms on the horizon of the broad, flat landscape at the edge of the ancient Niobrara River. In the dis-tance, a large herd of Meno-ceras, *a small rhinoceros, crosses the wide, many-chan-neled, sandy riverbed. The river's course is marked by a luxuriant gallery forest and the surrounding area is a green and light brown patchwork of trees, bushes, and grassy open-ings.* Dinohyus, *a giant pig-like entelodont, surprises a small herd of the gazelle camel* Stenomylus *that had been feeding on the tall grasses of the park woodland. A wooded patch of hackberry, oak, and locust partially conceals* Moro-pus, *a strange mammal that resembles a cross between a horse and a rhinoceros but has clawed feet. Groups of* Para-hippus, *the three-toed fore-bearer of modern horses, and the oreodont* Merycochoerus *feed in a clearing in the dis-tance. A slingshot horned antelope peers through a break in the forest.*

chapter NINE 9

Origin of the Modern World:

The Late Cenozoic

[Today]

The Agate Springs Quarries are nestled in the hills along the margins of the modern Niobrara River in the shortgrass prairie of western Nebraska. Meadowlark, antelope, deer, and coyotes live on the prairie where 20 million years ago the landscape was teeming with mammals adapted to a park woodland. The light gray beds of the Harrison Formation are evidence of the braided stream channel of the ancient Niobrara River and of a time when volcanic eruptions in Wyoming and Idaho belched white ash that choked the river with sediment. The climatic conditions were much drier and more seasonal than during the earlier Tertiary when tropical forests blanketed much of the American West.

World-famous quarries at Agate Springs, the site of Agate Springs Fossil Beds National Monument, preserve the dispersed and disarticulated carcasses of more than a hundred rhinoceroses. Bones from both adult and young rhinos of the species *Menoceras* are jam-packed into two layers. This small rhino, about the size of a modern tapir, had two small horns on the tip of its nose. The bone bedlam of the quarry suggests that the animals died during a summer drought. Although the bone bed is almost exclusively composed of rhino bones, two other animals, the odd *Moropus* and the formidable *Dinohyus*, are also known from skeletons in the area. *Moropus*, a perissodactyl related to the horse, was the size of a Clydesdale horse and possessed a tapir-like nose. Its body resembled a rhino but it had strange clawed feet and a head similar to a horse. Paleontologists speculate that it could have stood up on its hind legs to reach into the branches of trees to browse on the leafy vegetation. *Dinohyus*, the last of a long lineage of North American pig-like animals called entelodonts, may have been the most vicious predator of the Agate Springs landscape. Standing more than 6 feet at the shoulders, *Dinohyus* had a fearsome gaze and was fleet-footed. The skull of *Dinohyus* was massive—not that much smaller than the skull of *Tyrannosaurus rex*. Crush marks on the bones of *Moropus* demonstrate that *Dinohyus* scavenged or preyed upon larger mammals in the streamside gallery forests.

Near the rhino quarries are areas that have produced many fine skeletons of small camels and the intact dens and skeletons of bear-dogs. Ecologically resembling modern African gazelles, the graceful gazelle camel *Stenomylus* lived in herds and fed on grasses and herbs in open and wooded areas. The bear dogs were probably the hyenas of the ecosystem, killing animals and scavenging their carcasses. Other fossils from Agate include the long-necked camel *Oxydactylus*, the strange horned *Syndyoceros*, and a host of primitive pig-like oreodonts such as *Merycochoerus*. The land beaver *Paleocastor* is found in the bottoms of large corkscrew-shaped burrows. At first these burrows baffled paleontologists who named them *Daemonelix*, the Devil's Corkscrew.

The vertebrate fossils at Agate Springs are world-class, but fossil plant evidence for the nature of the environment is unusually hard to come by. The chemistry of the Harrison Formation was too alkaline and has resulted in the destruction of almost all of the plant material. The only fossils that have survived are the casts of roots and the tough seeds of the grass *Berriochloa* and of the woodland hackberry tree *Celtis*.

Hackberry seeds

[*Celtis* sp.]

The stony seeds of the hackberry trees and occasional grass seeds are the only common fossil plants in early Miocene rocks of Nebraska.

Early Miocene, 18 mya
Marsland Formation, Nebraska
Average seed diameter: 0.2 inch, DMNH 5123

Gazelle camels

[*Stenomylus hitchcocki*]

Small gazelle-like camels were some of the earliest mammals to have high crowned teeth. Plant remains found in the plaque on the teeth of these camels provide evidence that they fed on grass.

Early Miocene, 20 mya
Harrison Formation, Nebraska
Average skeleton length: 5 feet, DMNH 1204

chapter NINE

origin of the Modern world: The Late Cenozoic

The Miocene World

[20 MILLION YEARS AGO]

An ice cap forms at the South Pole, the climate cools, and the continents are drifting toward their present locations.

The modern world and its inhabitants

The past 24 million years have seen some of the most dramatic climatic changes in earth's history. During this time the Drake Passage between South America and Antarctica opened, creating the cold circum-Antarctic current. Huge mountain belts thrust up, forming the Alps, Himalayas, and Cascades. These high areas locally lowered temperatures and regionally caused rain shadows, drying the continental interiors. Massive continental ice sheets formed first on Antarctica and then later at high northern latitudes. The climate cooled and eventually led to the Ice Age 2 million years ago. The warm world of the early Tertiary became the planet that we inhabit today: cold at the poles and warm at the equator. A direct result of this climatic deterioration was the fragmentation of the worldwide forests of the early Tertiary and their replacement with more open habitats such as park woodlands, savannas, grasslands, and deserts.

Today, much of the central portion of North America is filled with a sea of grasses. Trees occur only near more permanent rivers, streams, and lakes. Grasslands, so important to humans and to grazing animals, are a relatively recent phenomenon. The oldest grass fossils in North America date from the Eocene, but grasses did not become common until well into the Miocene, less than 20 million years ago. The fossil record of grasses and recent work on carbon isotopes record the nearly simultaneous worldwide appearance of a new type of grass photosynthesis in North America, Africa, and Asia in the late Miocene, about 6 or 7 million years ago. A major mammalian extinction occurred at approximately this time. Browsing mammals were decimated, while the grazing mammals survived almost intact. Some researchers suggest that the grasslands we see today are extremely young, first occurring in western North America in the late Pleistocene, less than 1 million years ago.

New adaptations appeared among the mammals as habitats opened. Features that could be used to visually identify members of one's own species in an open area became much

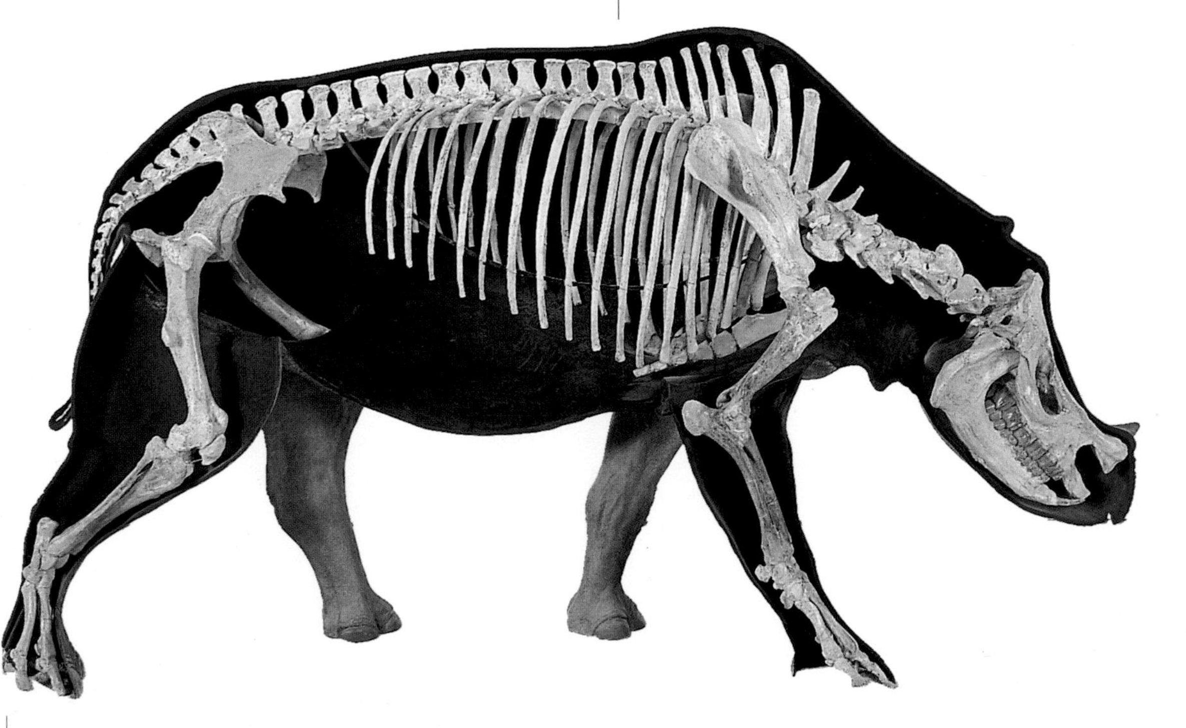

Two-horned rhinoceros

[Menoceras arikarense]

Rhinoceroses were common grazing animals in the Miocene savannas of the American West.

Early Miocene, 20 mya
Harrison Formation, Nebraska
Skeleton length: 6.2 feet, DMNH 129

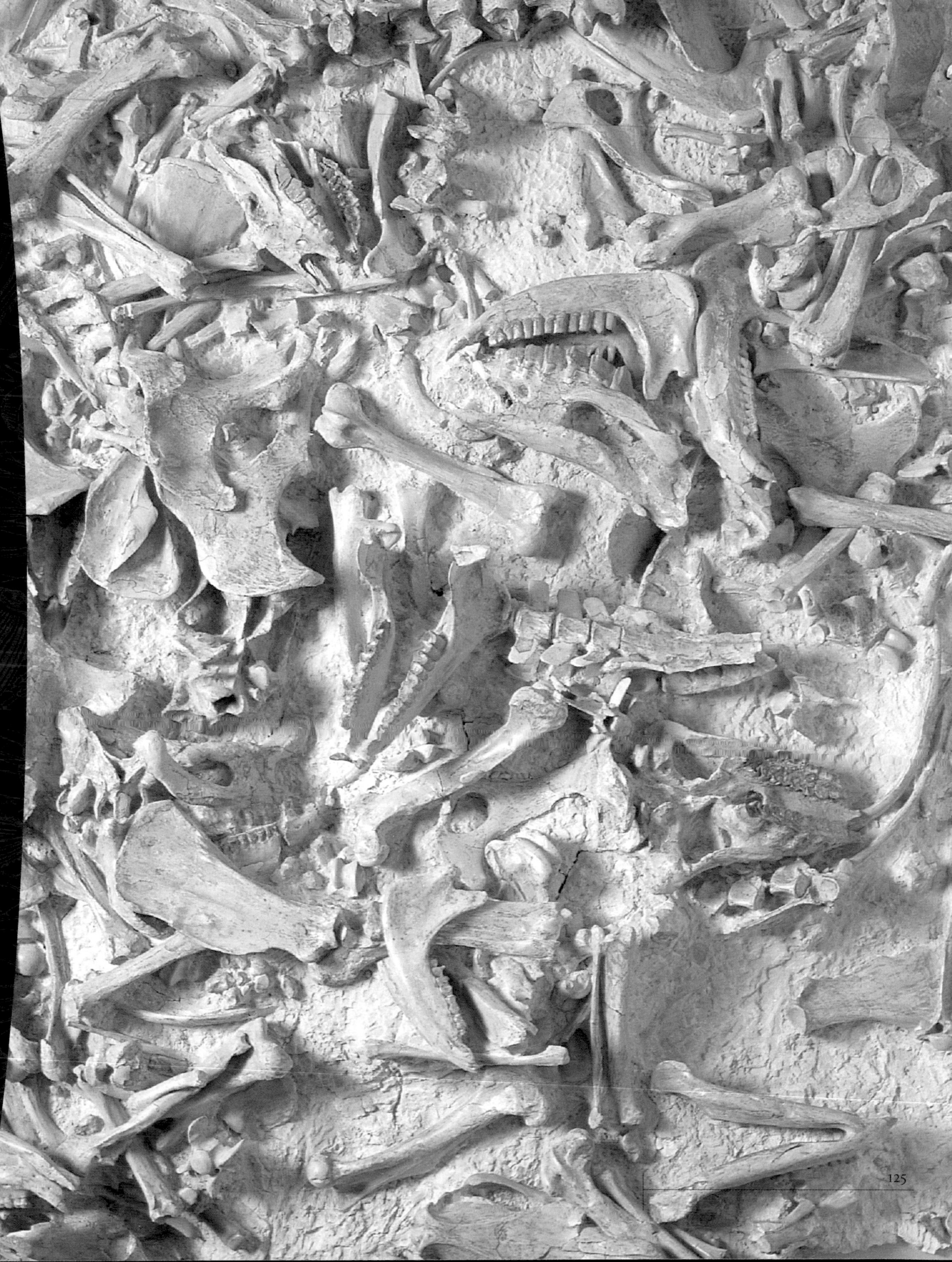

Rhinoceros bone bed

Twenty million years ago a herd of the small Menoceras *was preserved in the deposits of lakes that were in the floodplain of the ancient Niobrara River of western Nebraska. These quarry sites preserve in exquisite detail the bony anatomy of the skeleton.*

Rhinoceros bone bed

[*Menoceras arikarense*]

Early Miocene, 20 mya
Harrison Formation, Nebraska
Slab length: 6.2 feet, DMNH 129A

Chalicothere

Looking like a cross between a large horse and a rhinoceros, the chalicotheres were unusual odd-toed ungulates. They had claws rather than hooves on their feet, which enabled them to dig and split open plant materials for feeding. Scars on the bones of Moropus *match the proportions of the* Dinohyus *teeth, indicating that the* Moropus *was the prey of the giant pig-like entelodonts.*

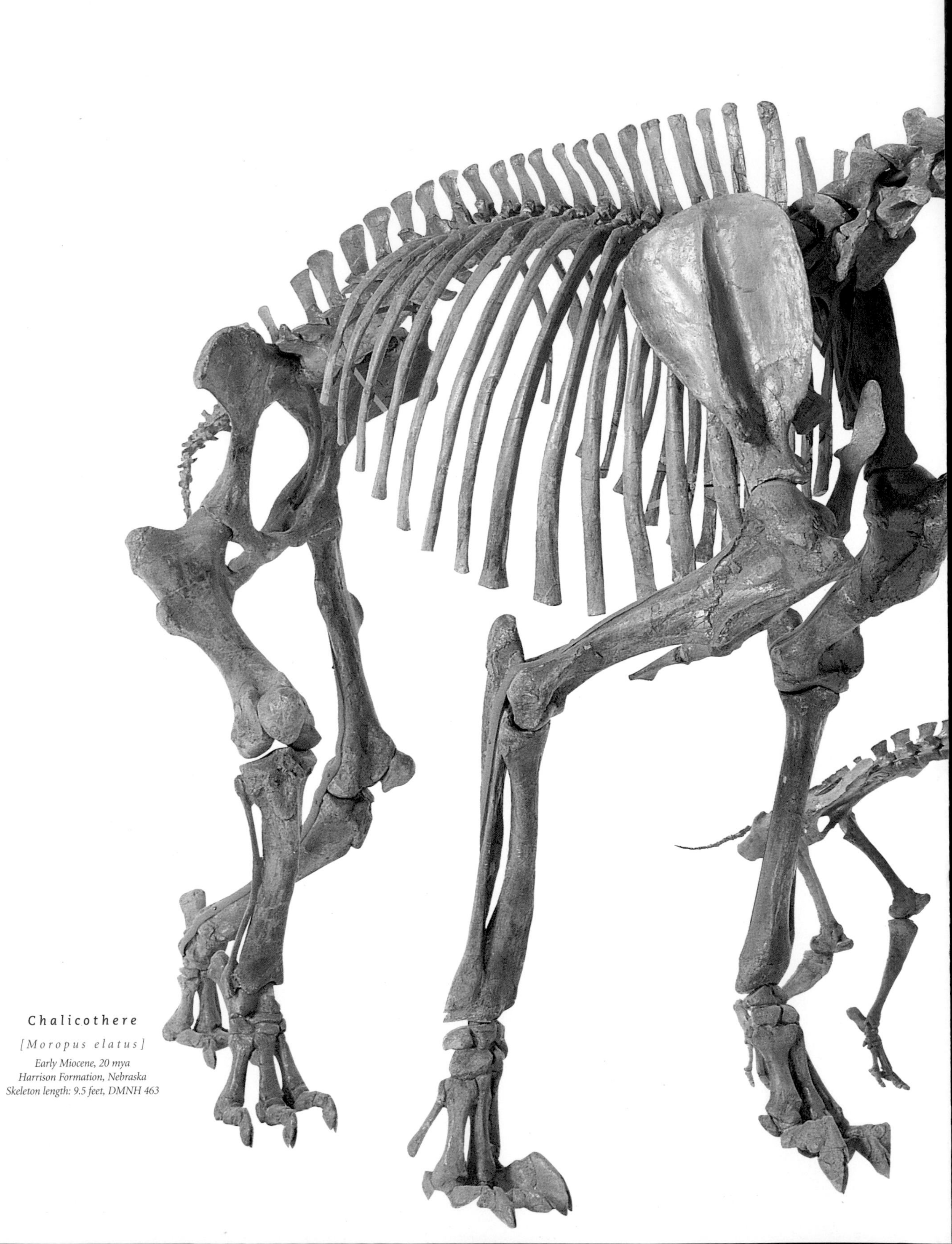

Chalicothere

[Moropus elatus]

Early Miocene, 20 mya
Harrison Formation, Nebraska
Skeleton length: 9.5 feet, DMNH 463

more common. Horns, antlers, and probably color patterns—all of which attracted mates—became much more elaborate, particularly among the deer, buffalo, and antelope families, which now dominated most mammalian communities. Even a Miocene rodent, *Ceratogaulus*, developed two horns on its nose.

As habitats opened, herding became more common. Evidence of herding behavior among late Tertiary mammals is found at quarry sites similar to Agate Springs. Large numbers of skeletons of one species of rhinoceros, horse, or artiodactyl are found buried together. The mass death sites of migrating caribou in Canada and wildebeest on Africa's Serengeti Plain are modern examples of herds that have the potential to become fossil bone beds.

Body size is another indicator of open habitats. Some species in almost all groups of mammals became larger in the late Tertiary. Elephants, sloths, hippos, rhinos, and even beavers developed to massive sizes. The largest terrestrial mammal of all time, *Paraceratherium* from Asia, reached 16 feet in height at the shoulder, exceeding the size of many dinosaurs. Body size seems to increase as a defense against predators because it is not possible to hide as well in an open habitat as in a closed one and size is a very successful defense strategy.

Perhaps also in response to predators, plant-eating animals in open areas developed adaptations for running. Among the horses, lateral toes were reduced and only the

Oreodont family

[Merycochoerus magnus]

Oreodonts may have lived in family groups. These skeletons of a male, female, and five newborns were found close together.

Early Miocene, 20 mya
Marsland Formation, Nebraska
Male skeleton (left) length: 4.9 feet
DMNH 1264
Female skeleton (right) length: 3.8 feet
DMNH 1263
Juveniles slab (above) length: 3 feet
DMNH 1265

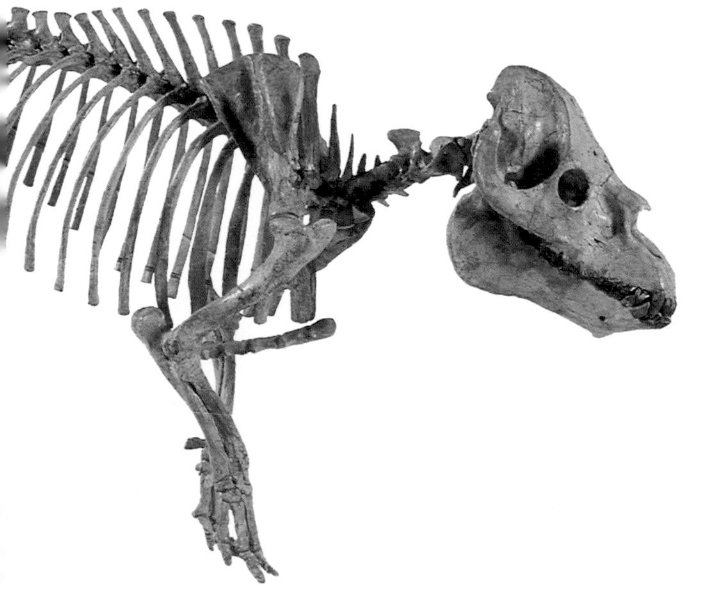

middle toe remained to support the animal. Among most groups of the more advanced artiodactyls, the toes were reduced to two middle digits fused into a single bone called the canon bone. Limb lengths increased for more efficiency and speed in outrunning predators.

As the habitats changed, a major shift in diet from softer leaf browse to coarse and gritty ground vegetation also occurred among many ungulates and other mammals. Tooth crowns became more complex with infolding of enamel and dentine in some rodents, perissodactyls, and artiodactyls. Higher crowned teeth also appeared in these groups and, as a consequence, the animals were able to process coarser vegetation without grinding their teeth away.

The fossil horses of North America are a good example of lineages of mammals that evolved to keep pace with an ever-changing environment. Horse evolution in North America began 55 million years ago with the appearance of *Hyracotherium*, which was large for its time but was really no larger than a small dog and much like one in overall body proportions. Its teeth were relatively simple in comparison to those of later horses, but they show an incipient ridged pattern like that found in later horses. The premolar teeth are simplified, not too different from those of its ancestors. *Hyracotherium* had four toes on its front foot, with its first digit lost and its fifth reduced. The hind foot bore three digits with the side toes extremely reduced in size. On both the front and hind limbs, the axis of support was through the middle digit as in all perissodactyls. As many as three species of *Hyracotherium* lived at any one time and perhaps a dozen existed until the next stage in horse evolution. *Orohippus* and *Epihippus* lived in the middle Eocene. They were very similar to *Hyracotherium* except that their premolar teeth had become more complex and similar to the molar teeth. Throughout horse evolution the major features of the horse skull remained remarkably similar.

During the latest Eocene and Oligocene, *Mesohippus* was the most common horse, although several other genera represent well-adapted side branches of the horse lineage. *Mesohippus* had more crested teeth and was slightly larger than its predecessors. It had three digits on both the front and hind limbs.

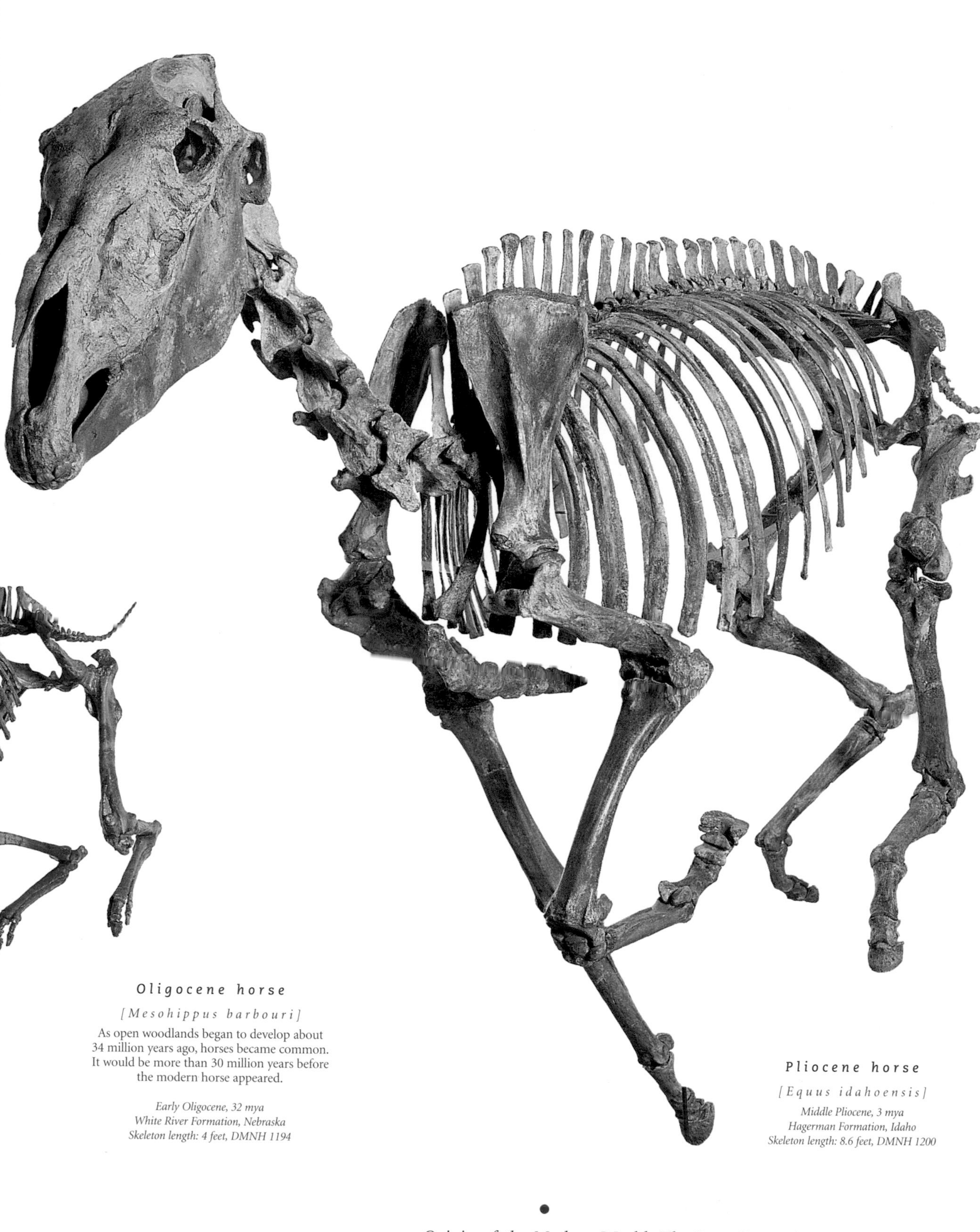

Oligocene horse

[Mesohippus barbouri]

As open woodlands began to develop about 34 million years ago, horses became common. It would be more than 30 million years before the modern horse appeared.

Early Oligocene, 32 mya
White River Formation, Nebraska
Skeleton length: 4 feet, DMNH 1194

Pliocene horse

[Equus idahoensis]

Middle Pliocene, 3 mya
Hagerman Formation, Idaho
Skeleton length: 8.6 feet, DMNH 1200

Horse evolution

The modern horse first evolved in North America and immigrated to Asia, Africa, and Europe before it became extinct here only 10,000 years ago. The horse was reintroduced into North America by the Spaniards in the 1500s.

Whale evolution

[RETURN TO THE SEA]

Whales evolved from long-jawed land mammals known as mesonychids. Fossils that record the transition from a terrestrial to a marine lifestyle are best preserved in Pakistan and Egypt. Eocene whales such as Pakicetus *and* Ambulocestus *had well-developed hind limbs. Later forms such as* Rodhocetus *and* Basilosaurus *retained legs but swam primarily with their broad tails. Fully marine whales, such as baleen and sperm whales and dolphins, evolved by the Miocene.*

Straight-tusked elephant

The elephants first immigrated from Asia to North America about 15 million years ago. They became extinct here only 12,000 years ago. The elephant family includes the gomphotheres, which had long tusks in both their upper and lower jaws.

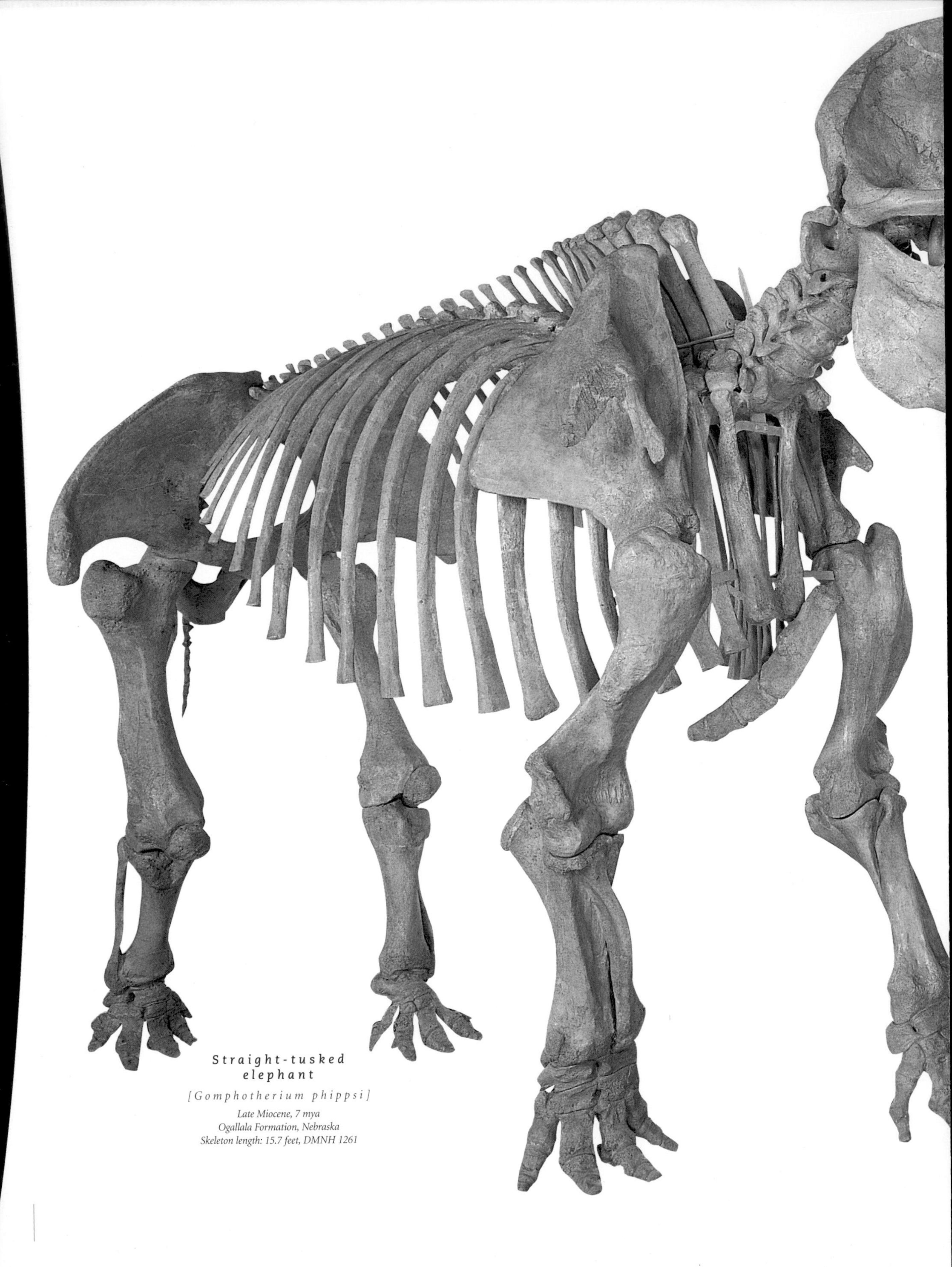

Straight-tusked elephant
[*Gomphotherium phippsi*]
Late Miocene, 7 mya
Ogallala Formation, Nebraska
Skeleton length: 15.7 feet, DMNH 1261

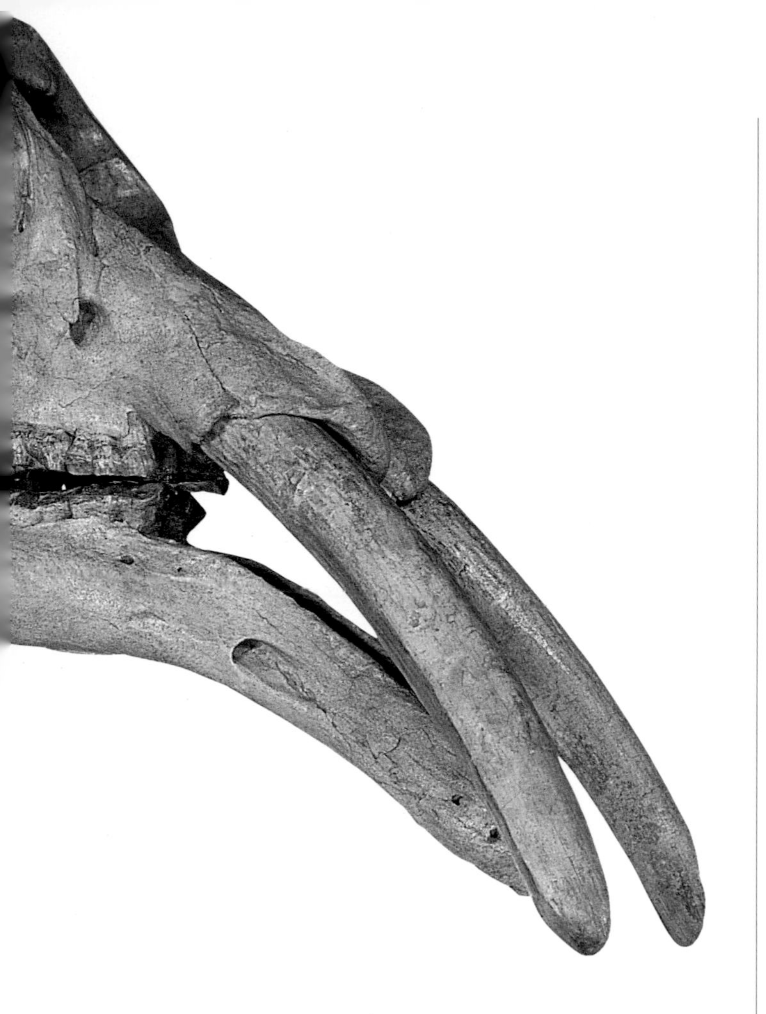

At least four separate lineages are known in the early Miocene. By the middle of the Miocene, horses had diversified considerably with more than 10 different lineages worldwide. Some of these lineages retained low-crowned teeth, but the major stem of horse evolution began to develop highly infolded and high-crowned molars that signaled their dietary change from browsing to grazing. *Equus*, the genus of living horses and zebras, finally appeared in the early Pliocene. It had a single digit, or hoof, on its front and hind feet, continuing the trend of side-toe reduction. The modern horse is superbly adapted for its habitat: a large animal that can eat grass and run like the wind. During the Pleistocene, between 5 and 15 lineages of *Equus* lived in North America.

The changing climate of the late Tertiary affected marine environments as well as terrestrial ones. As the polar waters grew colder, major oceanic currents such as the Gulf Stream and the Japanese current were initiated and the world's oceans began to mix. Movement of nutrients and upwelling currents increased the productivity of the oceans, and a whole new group of animals began to take advantage of the marine smorgasbord. Whales, which had appeared in the Eocene, diversified into the modern toothed and baleen-bearing species by the late Oligocene. By the Miocene, baleen whales had achieved the massive body size that allowed them to process tons of plankton a day. Other mammals also began to exploit the maritime larder. Carnivores took to the water and by the Miocene the first seals, sea lions, walruses, and sea otters were active predators of the coastlines.

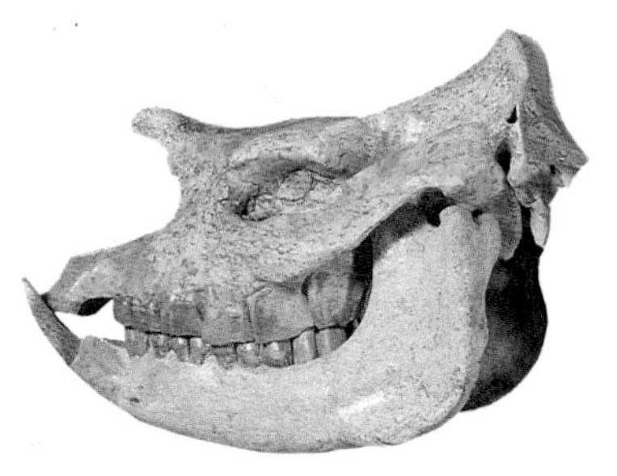

Rhinoceros skull

[Teleoceras hicksii]

Rhinoceroses were common animals of the late Tertiary grasslands of North America. This one had a very squat body shape, suggesting that it may have lived like a hippopotomus.

Late Miocene, 7 mya
Ogallala Formation, Colorado
Skul length: 2 feet, DMNH 304

As the Tertiary came to a close, the continents drifted into their modern positions and the world as we know it came into existence. Antarctica remained on the South Pole, and ice destroyed its once diverse terrestrial fauna and flora. Australia and New Guinea drifted north, colliding with the islands of Indonesia, but the deep oceanic trenches prevented the mixing of its predominantly marsupial fauna with the placental fauna of the Northern Hemisphere.

South America was like Australia, a southern island continent for much of the Tertiary; it had a large diversity of marsupial mammals as well as some bizarre placentals. But not all of the marsupials of South America were mild-mannered herbivores. Many were carnivorous. The dog-like borhyaenids

Lucy, an early hominid

The human ancestor Australopithecus *is first known from about 4 million years ago. Although it had a relatively small body, a small brain, and a grasping big toe, details of its skeleton and fossil footprints show that it walked upright.*

Lucy

[Australopithecus afarensis]

Late Pliocene, 3.2 mya
Hadar, Ethiopia
Skeleton height: 3.6 feet, DMNH 22980

and saber-toothed thylacosmilids competed with giant predatory flightless birds called phorusrhacids. The herbivorous fauna was composed of edentates (anteaters, sloths, and armadillos) and now extinct notoungulates. In a superb example of convergent evolution, the notoungulates of South America developed many of the same adaptations to an opening landscape that were evolved by unrelated plant-eating mammals in the Northern Hemisphere. Thus, we see horse-like, hippo-like, and camel-like notoungulates appearing in South America at more or less the same time we see horses and camels in North America and hippos in Africa.

Around 8 million years ago, North and South America became connected at the Isthmus of Panama, a land bridge that allowed for what is called the Great American Interchange. Since that time, mammals have migrated both north and south across the isthmus. North America received marsupials, such as the opossum, and edentates, such as the armadillo, and sent south a whole suite of animals, including camels, deer, horses, elephants, dogs, cats, and mink. The arrival of the North American immigrants coincided with and perhaps caused the demise of the carnivorous marsupials and notoungulates. Many of the animals we think of as South American (like llamas, tapirs, and jaguars) are actually descendants of North American immigrants of many thousands of years ago.

The human lineage

The opening of the terrestrial habitats in the late Tertiary in Africa caused one of the most significant biotic events in earth history: the evolution of humans. During the Miocene, the hominoids, or ape-like forebears of the modern hominids (modern apes and humans), were much more diverse than the monkeys in Africa. That relationship is reversed today. Ranging across the European, Asian, and African continents, early hominoid groups such as the dryopithecines and sivapithecines included as many as a dozen species. Among this group, several genera, which lived from 14 to 10 million years ago, had teeth similar to what we would expect for the ancestor of humans and apes. These primates lived an arboreal lifestyle in predominantly tropical habitats of Africa, Europe, and Asia.

Between 8 and 4 million years ago, the first human began to walk habitually on two legs, the major feature that marks the origin of humans. Evidence discovered in the 1980s and early 1990s shows that the earliest species in the human lineage, *Andropithecus ramidas*, walked erect but still resembled the apes in its dentition. A later species, *Australopithecus afarensis*, also walked erect, as footprints from Laetoli in East Africa demonstrate, and had grasping hands. Its foot still retained a grasping large toe, however, much like that in the African apes. This human showed strong sexual dimorphism; males were nearly 5 feet tall and females were only slightly more than 3 feet tall. These early humans lived in forested areas and may have been able to climb trees to forage and to sleep out of the reach of predators. There is no evidence of tool use, but tool use by living great apes suggests that simple tools made of wood twigs and rocks may have been within the early human repertoire.

Two million years ago, *Homo erectus* appeared and shortly thereafter immigrated out of Africa into Asia. Approximately 1 million years ago, humans reached Europe. By this time human brain size had nearly doubled and both males and females were more nearly the same body size. Tool use had achieved a high level of organization with symmetrical hand-axe forms, blades, and scrapers. Evidence suggests that prior to the advent of *Homo sapiens*, fire, a home base, and shelter had developed within the human culture.

The Ice Age

Dramatic fluctuations in climate have characterized the past 2 million years. The Pleistocene epoch is often called the Ice Age because glacial ice covered much of North America and western Eurasia. The Ice Age was really a series of ice ages separated by warm periods known as interglacial periods. During some of the interglacials, world climate was warmer than it is today. During the glacial periods, massive continental ice sheets lay on the continents. The size of these ice sheets was almost unimaginable. The most recent, in place a mere 18,000 years ago, covered the northern half of North America with a slab of ice more than 2 miles thick. Its southern margin formed a 500-foot-tall sheer ice cliff that

Oldowan pebble tool

[Tanzania]

This pebble tool was collected in Olduvai Gorge by Louis Leakey. It is the oldest direct evidence of tool use by hominids.

Late Pliocene, 1.75 mya
Olduvai Gorge, Tanzania
Length: 3.9 inches, DMNH A2013.1

Continental ice sheet

[Ellesmere Island]

The continental ice sheets on Antarctica, Greenland, and the Canadian Arctic islands are powerful reminders of what New York, Chicago, and Seattle looked like 14,000 years ago when glacial ice covered the northern half of North America.

Mammuthus columbi

The mammoth was hunted by the first humans to migrate into North America. Many mammoth skeletons have been found associated with artifacts of early Native Americans.

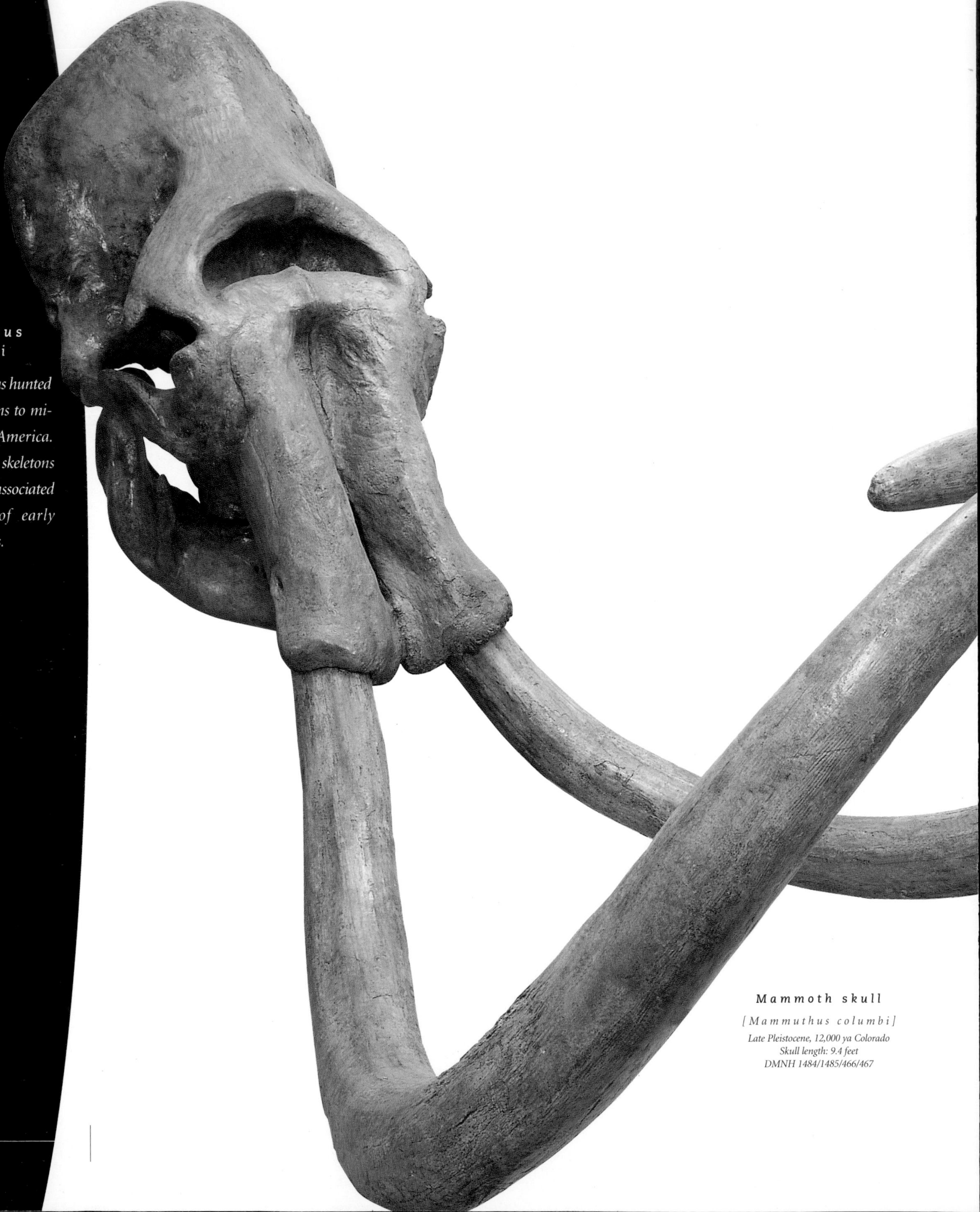

Mammoth skull

[*Mammuthus columbi*]

Late Pleistocene, 12,000 ya Colorado
Skull length: 9.4 feet
DMNH 1484/1485/466/467

extended from Seattle to Chicago to New York. So much water was locked into the continental ice of North America and Eurasia that world sea level was 300 feet lower than present. The rapid advance and retreat of these ice sheets during the past 2 million years threw the world into a series of rapid and dramatic ecological reorganizations. Near the edge of the massive ice sheets, vegetation zones expanded and contracted with the ice. Far from the ice, the rapidly changing climate still had dramatic effects. Fossil pollen from South America indicates that the tropical rainforests were repeatedly replaced with tropical savannas as the ice sheets waxed and waned. Fossil hippopotamuses and crocodiles from sites in the Sahara Desert also show that the amount of rainfall in tropical zones has varied with the fluctuations of ice.

Until 12,000 years ago, a diverse large mammal fauna occupied much of the North American continent. Large carnivores, such as the saber-toothed cat, lion, dire wolf, and giant short-faced bear, preyed upon camel, horse, elk, deer, antelope, bison, and moose. Musk ox, typical of modern Arctic regions, lived in the midcontinent, as did now extinct South American immigrants like the giant ground sloths, which extended their ranges northward. The giant beaver was about 8 feet in length and probably weighed more than 300 pounds. *Mammuthus*, the giant mammoth of the Pleistocene, was 16 feet tall. It first arrived in North America from Eurasia about 1.9 million years ago. A common element in most Pleistocene faunas, the mammoth was an open-habitat animal whereas its relative, the mastodon, probably inhabited forested regions. The rapid climate variations of the Pleistocene resulted in a variety of strange occurrences: cheetahs in Wyoming, small mammoths in California, lemmings in Colorado, hippopotamuses in London, and walruses in New Jersey. It is only because of the recent warming and cooling of the planet that the occurrences of these animals seem odd to us.

The first humans crossed the Bering land bridge from Siberia into North America earlier than 12,000 years ago and perhaps as long ago as 25,000 years. The time of arrival coincides with the extinction of much of the Pleistocene megafauna. Sometime around 12,000 to 11,000 years ago, the mammoths, mastodons, horses, camels, giant sloths, and dozens of other large mammals became extinct in North America. Although the cause of this recent extinction is a mystery, human hunters and habitat change associated with cooling and warming cycles are the two most often cited suspects. Given that the average duration of interglacials was as long as 100,000 years, it is possible that we are actually living in an interglacial and that the ice sheets will return sometime in the not-too-distant future.

The future

We live in a world that will continue to change as it has in the past. The fossil record indicates a series of steps from the simplest of life forms to more complex ones. The planet is transformed by massive but slow changes in the positions of the continents and the nature of the climates that govern the flora and fauna that live on them. Faunas and floras change in response to ecological interactions.

Humans are now the dominant element in almost all habitats. We have developed the technology to live in the most frigid and the most arid regions of the globe. We are continually modifying the communities within which we live. At the current rate of population increase, there will be more than 10 billion humans by the middle of the next century. With increasing population, increased resource use is a given. More energy, more food, and more space will be required to sustain the increase.

The human impact on our natural world is occurring at a much greater rate than at any previous time in earth's history. It rivals and often exceeds the damage done by the largest catastrophes. From paleontology, we learn that it has taken 3.5 billion years to develop life to its present state. We have a responsibility to ensure that our children continue to enjoy the earth's vast natural beauties and that the descendants of other life forms flourish well into the future.

Acknowledgments

Prehistoric Journey, the exhibition at the Denver Museum of Natural History, was made possible by the concerted and dedicated efforts of many individuals between 1989 and 1995.

PREHISTORIC
JOURNEY℠

Management Team
Brian McLaren, Richard Stucky, Rebecca Smith, and Merry Dooley

Interpretive Development Team
Frances Kruger, Kirk Johnson, and Rebecca Smith

Exhibit and Graphic Design
Robin Farrington, Allison Sundine, Bryce Snellgrove, Charles Stirum, Mike Coates, Carl Shipley, Scott Brown, Julie Langenthal, Lisa McGuire, Lexie Foster, Jim Marlow, Jack Augenblick, and Jim Leggitt

Fabrication and Carpentry
Bryan Bardwell, Dave Burdekin, Kevin Johnson, Rob Jurado, Marty Lechuga, Fred Jurado, Brian Wheeler, Victor Muñoz, John Dilgarde, Jeff Yearick, Ken Portuese, and Dana Seymour

Lighting and Special Effects
Charles Stirum, Daniel Gonzales, and Zachary Reynolds

Videos and Computer Animation
David Baysinger, Eric Klepinger, Elizabeth Gilmore, and Terry Trieu

Fossil Preparation
Kenneth Carpenter, Bryan Small, Jon Christians, Jennifer Moerman, Jerry Harris, Karen Alf, Linda Smith, and Bill May

Illustrations
Greg Michaels, Marjorie Leggitt, and Tom Buchanan

Photography
Nancy Jenkins, Rick Wicker, David Shrader, Gary Hall, Dave McGrath, Richard Stum, and Beth Kaminsky

Administration
Libbie Gottschalk, Marianne Reynolds, Sheila Mutchler, Alan Espenlaub, Curt Simmons, Karen Arnedo, and Patricia Jablonsky

Interpretation
Frances Kruger, James Alton, and Judy Schaefer

Evaluation
Margaret Marino and Elaine Owens

Sculpture
Tom Shankster, Dennis McElvain, Gary Staab, Mark Akerley, and John Gurche

Curation and Scientific Content
Kirk Johnson, Richard Stucky, Kenneth Carpenter, Tom Hardy, Logan Ivy, Jennifer Snyder, and Elise Schloeder

Foregrounds
Elizabeth Marshall, Hugh Watson, Denise Patton, Leonard Koenig, Jayme Irvin, and Claudette Wallace

Murals
Jan Vriesen, Deborah Vriesen, Kent Pendleton, Terry Chase, Jeff Wrona, Chuck Parson, and Tom Buchanan

Education
Rebecca Smith, Merry Dooley, Eddie Goldstein, Kathy Kuba, Jim Goddard, and Jeff Stephenson

Development
Ann Fothergill and Bonnie Downing

Education Advisory Committee

Thom Adorney, Diana Scheidle Bartos, Ann Bell, Marsha Barber, Steve Gigliotti, Kay Bard Gray, Glen McGlathery, Pam Schmidt, Nancy Songer, and Sharon Stroud

Scientific Consultants

Allison Palmer, Reuben Ross, William Cobban, Leo Hickey, William DiMichele, Howard Feldman, Gene and Royal Mapes, Don and Joanne Mikulic, Hermann Pfefferkorn, Sidney Ash, Douglas Nichols, Joseph Thomasson, Lynn Margulis, Sergius Mamay, Scott Wing, Garland Upchurch, Cathy Whitlock, Leonard Krishtalka, Robert Hunt, and Malcolm McKenna

Special thanks are due to the many talented and dedicated volunteers without whom the project would not have been possible.

The exhibit was made possible by major donations from the Schlessman Family Foundation, Amoco Foundation, National Science Foundation, Boettcher Foundation, Gates Foundation, El Pomar Foundation, Martin J. and Mary Anne O'Fallon Trust, Helen K. and Arthur E. Johnson Foundation, Denver Foundation, Pauline A. and George R. Morrison Charitable Trust, Friends of Allan R. Phipps, and Friends of James W. Vanderbeek.

Additional sponsors include Marcy and Bruce Benson, Philippe Dunoyer, Members of the Museum, Concord Services, Inc., Martin Marietta Astronautics Group, Oakley Industries, Inc., Gorsuch Kirgis L.L.C., the family of Henry C. Van Schaack, and Oren and Beverly Benton.

Book credits

Illustrations

Cover painting by John Gurche; chapter opening paintings by Terry Chase (Chapters 2 and 3), Deborah Vriesen (Chapter 4), Jan Vriesen (Chapter 5), Jeff Wrona (Chapters 6 and 9 [composite]), and Kent Pendleton (Chapters 7 and 8); paintings by Greg Michaels on pages 17, 18, 20, 23 bottom, 25, 34, 45, 53, 56, 59, 70, 85, 87, 90, 116, 131; paleogeographic reconstructions based on maps by Christopher Scotese, illustrated by Ann Douden; and fossil pattern art by Marjorie Leggitt.

Photography

All photographs by Rick Wicker except those by David McGrath (pages 27 bottom, 33, 36, 39 top, 39 bottom, 73, 77 top, 93, 99 bottom), Kirk Johnson (pages 8, 14, 30, 50, 82, 122, 135 bottom), Catherine Joy Johnson (page 42), Richard Stum (page 43 lower right), Gary Hall (pages 83 left, 95 bottom, 106), Richard Stucky (page 115 lower right), Nancy Jenkins (page 104), Hal Stoelze (page 95 top), and Dick Deitrich (page 64).

Specimens

Specimens in this book are largely from the Denver Museum of Natural History collections. People who donated specimens that are used include: Sidney Ash, Jeff Carpenter and Collette Cherpin, Richard Dayvault, Howard and Darlene Emry, Bob Farrar, Val and Lloyd Gunther, Bill Hawes, Bill Higbee, Ron Horst, and Paula Ott, Mario Hünicken, Kirk Johnson, Ruth Johnson, Neal Larson, Peter Larson, Royal Mapes, Ron Meyer, Allison Palmer, Tony Pfeiffer, Mr. Pohndorf, Wayne Smiglewski, Glenn Stewart, Steven Tuftin, Rene Vandervelde, Chris Weege, and Wesley Wehr. People who collected specimens that are illustrated in this book include Karen Alf, Harley Armstrong, Peter Bucknam, Harold Cook, David Craig, Frank Figgins, J. D. Figgins, Russell Hendee, Jeremy Hooker, B. F. Howarter, Kirk Johnson, Robert Landburg, Don Lindsay, George Lonford, Harvey Markman, Matthew McLaren, Marshall McLean, Jennifer Moerman, Pablo Puerta, James Quinn, Philip Reinheimer, Pam Schmidt, Din Seaver, Morris Skinner, Bryan Small, Richard Stucky, Allison Sundine, Garland Upchurch, Nelson Vaughan, Kevin Werth, and India Wood.

Land access was provided by the Bureau of Land Management, the National Forest Service, and Union Pacific Resources.

Institutions that loaned or exchanged specimens illustrated in this book include the Field Museum of Natural History, Harvard Museum of Comparative Zoology, U.S. Geological Survey, Western Wyoming Community College, Carnegie Museum of Natural History, University of Michigan, Smithsonian Institution, and American Museum of Natural History.

Glossary

aetosaurs armored reptiles, known primarily from the Triassic period

ammonites extinct mollusks with coiled and straight shells; relatives of modern octopus, squid, and nautilus

amniotes animals such as birds, mammals, and reptiles whose embryos are enclosed in a desiccation-resistant membrane

amphibians tetrapods that do not produce an amniotic egg and can live both on land and in the water

anapsid skull of a reptile which has no muscle openings; found in turtles

angiosperms plants that produce flowers and fruits

ankylosaurs: armored ornithischian dinosaurs, known from the Jurassic and Cretaceous periods

anoxia condition resulting from a lack of oxygen

arthropods invertebrate animals with external skeletons and paired appendages (e.g., insects, trilobites, chelicerates, crustaceans)

artiodactyls even-toed ungulates (e.g., deer, antelope, bison)

asexual cell division reproduction by splitting of a cell

baculites straight-shelled ammonites, known from the Cretaceous period

bivalves invertebrates with paired shells (e.g., clams, oysters)

borhyaenids carnivorous marsupials that lived in South America during the Tertiary period

brachiopods invertebrates with paired shells, common during the Paleozoic era

bryozoans simple animals that form lace-like colonies of hundreds of individuals

burrow trace a hole in the sediment, often filled in with mud, made by a burrowing animal

calamites stems of sphenopsid plants

callipterids feathery-leafed seed ferns that were common during the late Paleozoic era

carbon-isotope record the variation in the relative amounts of the isotopes of carbon; useful for understanding ancient climate and biochemistry

carnivores flesh-eaters

cephalopods swimming mollusks with tentacles (e.g., squid, nautiluses, ammonites)

ceratopsians horned ornithischian dinosaurs known primarily from the Cretaceous period (e.g., *Triceratops*)

champsosaurs crocodile-like reptiles; common during the Cretaceous and early Tertiary periods

chelicerates arthropods that have a pair of limbs in front of their mouth (e.g., spiders, scorpions, eurypterids)

chert rock composed of silica that forms chemically in sediments

chloroplast organelles cellular structures where photosynthesis takes place

compression fossil a fossil formed by the flattening of an organism during burial

conifers trees with needles or blade-like leaves and seed-bearing cones

cordaites extinct trees related to conifers; cordaites have strap-like leaves and seed-bearing cones

creodonts carnivorous mammals that lived during the Eocene epoch

crinoids echinoderms with a hard skeleton; because they look like flowers, they are often called sea lilies

cuticle the cutinous outer coating found on arthropods; the waxy outer coating on leaves

cyanobacteria primitive single-celled organisms with no cell nucleus

cycadophytes the group of plants that includes cycads and their extinct relatives, the cycadeoids, or bennititaleans

diapsid skull with two openings for muscle attachment; found in dinosaurs, crocodiles, lizards, and snakes

dromeosaurs small, aggressive carnivorous dinosaurs that lived during the Cretaceous period

dryopithecines ape-like primates that lived during the Miocene epoch

echinoderms invertebrates with external skeletons and five-fold symmetry (e.g., crinoids, sea urchins, sand dollars)

edaphosaurs a type of pelycosaur or protomammal, some of which had a sail-like fin running the length of the spine

edentates an order of mammals that includes armadillos, anteaters, and sloths

entelodonts a family of extinct, pig-like artiodactyl mammals

estuary the portion of a coastal river that is affected by tidal activity

eukaryotes organisms that have cells with nuclei; includes all plants, animals, fungi, and some algae

eurypterids chelicerate marine and aquatic arthropods; eurypterids failed to survive the Permian-Triassic extinction

exaptation utilization of a trait for a purpose different from that which it evolved

fossil evidence of ancient life preserved in the geologic record

gigantopterids enigmatic seed plants with giant leaves; known only from the Permian period

graptolites lacy, floating marine organisms; common in the Paleozoic era

greenhouse effect a rise in global temperatures caused by the accumulation of insulating gasses in the earth's upper atmosphere

gymnosperms seed plants that do not produce angiosperm flowers

hadrosaurs duck-billed ornithischian dinosaurs that lived during the Cretaceous period

herbivores plant-eaters

holdfasts structures that attach crinoids to the substrate

ichthyosaurs marine reptiles of the Jurassic and Cretaceous periods

inoceramids a type of clam common in the Mesozoic era

insectivores insect-eaters

isotopes chemically similar atoms that differ in weight and stability

lycopods plants with scale-like leaves and spore-bearing cones; modern lycopods are herbaceous, but some Paleozoic lycopods were large trees

magma molten rock

mantle the portion of the earth that underlies the crust and extends to a depth of 1800 miles

marsupials mammals that carry their young in a pouch (e.g., kangaroos, koalas)

mastodons large browsing elephants; common during the ice ages

mesonychids a group of primitive, carnivorous mammals that are the ancestral stock of whales

metamorphic rocks rocks that have been compressed or heated but not fully melted

metazoans multicellular animals

metoposaurs large-headed amphibians that were common in the Triassic period

mollusks invertebrates with a large muscular projection called a foot (e.g., clams, snails, cephalopods)

monotremes egg-laying mammals (e.g., platypuses, echidnas)

mudstone rock composed of lithified mud

multituberculates extinct mouse-sized mammals with very elaborate teeth

nodosaurs a group of armored ornithischian dinosaurs related to ankylosaurs

nothosaurs a group of small marine reptiles related to plesiosaurs

notoungulates a type of hoofed mammals common in South America during the Tertiary period

omnivores animals that eat a varied diet including flesh, plants, and insects

ornithischian one of the two major groups of dinosaurs, characterized by a bird-like pelvic structure

osteolepiforms a group of primitive, lobe-finned fish that may be ancestral to amphibians, reptiles, birds, and mammals

ostracods arthropods with paired shells that look like clams

pachycephalosaurs ornithischian dinosaurs with a thick bony skull

paleontology the study of fossils

palynology the study of fossil pollen, spores, and other plant microfossils

pelycosaurs a group of late Paleozoic protomammals

perissodactyls odd-toed, ungulate mammals (e.g., horses, rhinoceroses, tapirs)

permafrost permanently frozen soil

petrifaction the process of replacement of original material by a water-borne mineral such as silica

phorusrhacids carnivorous flightless birds that were common in South America during the Tertiary period

photosynthesis a chemical process during which plants form carbohydrates and oxygen from sunlight and carbon dioxide

phylum major category of life forms; subdivisions of Kingdoms

phytosaurs large crocodile-like reptiles that were common during the Triassic period

placoderms bony-headed, jawed fishes

plankton microscopic marine plants and animals

plate tectonics the study of the motion of the large plates that compose the earth's crust

plesiosaurs large marine reptiles with paddle-like flippers that were common in the Jurassic and Cretaceous periods

progymnosperms extinct spore-bearing plants with conifer-like wood and fern-like leaves

prokaryotes single-celled organisms with no nucleus

protocetes the first whales

protomammals the group of animals, formerly known as mammal-like reptiles, that gave rise to mammals

pteridosperms an extinct group of gymnosperm plants that had fern-like foliage but bore seeds instead of spores

pterosaurs flying reptiles that lived during the Mesozoic period

rudists giant reef-building clams whose shells were extremely thick and cone-shaped

sandstone rock formed from cemented sand

saurischians one of the two major groups of dinosaurs characterized by a lizard-like pelvic structure

sauropodomorphs large herbivorous saurischian dinosaurs (includes sauropods and prosauropods)

sauropods large herbivorous saurischian dinosaurs with long necks and long tails; common in the Jurassic and Cretaceous periods

savanna a landscape characterized by open patches of trees and low herbaceous plants (usually grasses)

sedimentary rocks rocks formed by the accumulation of mineral grains, organic matter, or chemical precipitates

shale mudstone that has distinct layering

sivapithecines a group of ape-like hominid primates; known from the Miocene epoch

sphenodontids a primitive group of reptiles that are closely related to lizards; the only living example is the New Zealand tuatara

sphenopsids plants with whorled leaves, spore-bearing cones, and segmented stems (e.g., horsetail rushes)

sporangia small structures that bear the spores on a spore-bearing plant

spring and neap cycle monthly cycle of the waxing and waning of the tides

stegosaurs herbivorous ornithischian dinosaurs characterized by large dorsal plates, a spiked tail, and a very small head

stigmarian system the bifurcating root system unique to lycopod trees

stromatolites dome-like structures formed when sticky organic filaments of algae trap mud

synapsids animals that have a single skull opening for muscle attachment; includes both protomammals and mammals

taeniopterids a group of plants, probably related to cycads, that produced large, simple leaves with fine parallel veins

tetrapods animals that have four limbs (e.g., amphibians, reptiles, mammals)

thecodonts primitive reptiles that were the ancestral stock for dinosaurs, pterosaurs, and crocodilians

therapsids advanced protomammals, common during the Permian and Triassic periods

theropods carnivorous saurischian dinosaurs, including the smaller coelurosaurs and the larger carnosaurs

thylacosmilids saber-toothed marsupial carnivores that lived in South America in the late Tertiary period

trigontarbids small terrestrial arthropods

trilobites extinct arthropods with bodies divided into three segments

troodontids small carnivorous saurischian dinosaurs

uintatheres large, extinct herbivorous mammals with saber-like canine teeth and six horns

ungulates hoofed mammals

velociraptors small saurischian dinosaurs with large claws on both their hands and feet

selected Bibliography

Bakker, Robert T. *The Dinosaur Heresies.* New York: William Morrow and Co., 1986.

Behrensmeyer, A.K., J.D. Damuth, W.A. DiMichele, R. Potts, H.D. Sues, and S.L. Wing, eds. *Terrestrial Ecosytems Through Time.* Chicago: University of Chicago Press, 1992.

Briggs, D.E.G., and P.R. Crowther, eds. *Palaeobiology: A Synthesis.* Oxford: Blackwell Scientific Publications, 1990.

Carroll, Robert L. *Vertebrate Paleontology and Evolution.* New York: W.H. Freeman and Co., 1988.

Clarkson, E.N.K. *Invertebrate Palaeontology and Evolution.* 3rd ed. London: Chapman and Hall, 1993.

Gould, Stephen J. *Wonderful Life: The Burgess Shale and the Nature of History.* New York: W. W. Norton and Co., 1989.

Gould, Stephen J., ed. *The Book of Life: An Illustrated History of the Evolution of Life on Earth.* New York: W. W. Norton and Co., 1993.

Margulis, Lynn, and Dorian Sagan. *What Is Life?* New York: Simon and Schuster, 1995.

Margulis, Lynn, and Karlene V. Schwartz. *Five Kingdoms: An Illustrated Guide to the Phyla of Life on Earth.* New York: W. H. Freeman and Co., 1988.

Psihoyos, Louie, with John Knoebber. *Hunting Dinosaurs.* New York: Random House, 1994.

Schopf, J. William. *Major Events in the History of Life.* Boston: Jones and Bartlett Publishers, 1992.

Stewart, Wilson N., and Gar W. Rothwell. *Paleobotany and the Evolution of Plants.* 2nd ed. Cambridge: Cambridge University Press, 1993.

Ward, P.D. *The End of Evolution.* New York: Bantam Books, 1994.

Weishampel, David B., Peter Dodson, and Halszka Osmólska. *The Dinosauria.* Berkeley: University of California Press, 1990.

Index